Productivity in Construction Projects

Scrivener Publishing
100 Cummings Center, Suite 541J
Beverly, MA 01915-6106

Publishers at Scrivener
Martin Scrivener (martin@scrivenerpublishing.com)
Phillip Carmical (pcarmical@scrivenerpublishing.com)

Productivity in Construction Projects

Ted Trauner
Chris Kay
and
Brian Furniss

WILEY

This edition first published 2022 by John Wiley & Sons, Inc., 111 River Street, Hoboken, NJ 07030, USA
and Scrivener Publishing LLC, 100 Cummings Center, Suite 541J, Beverly, MA 01915, USA

For more information about Scrivener publications please visit www.scrivenerpublishing.com.

Wiley Global Headquarters
111 River Street, Hoboken, NJ 07030, USA

For details of our global editorial offices, customer services, and more information about Wiley products visit us at www.wiley.com.

Library of Congress Cataloging-in-Publication Data

ISBN 978-1-119-91080-0

Cover image: Pixabay.Com
Cover design by Russell Richardson

Set in size of 11pt and Minion Pro by Manila Typesetting Company, Makati, Philippines

Printed in the USA

10 9 8 7 6 5 4 3 2 1

Contents

Acknowledgments

As noted in the Forward to this book, many people contributed to the effort shown herein. Special thanks go to the following: William Manginelli, Scott Lowe, Mark Nagata, John Crane, Bruce Ficken, Richard Burnham, Stephanie Trauner, Natalie Furniss, Richard Browne, Michael Furbush and Janet Montgomery. Hopefully, we included everyone.

1
Purpose

Why should anyone read a book on productivity, particularly one about productivity in construction? For many people and entities in the construction business, productivity probably hasn't been of primary importance in the scheduling and execution of the work. But in today's construction environment, productivity should be a primary focus for everyone's future success.

Without a doubt, the construction industry is challenged today more than ever before. Many new construction opportunities exist, yet the industry has a dearth of subcontractors to rely upon to accomplish much of the work, and this was before skilled labor shortages and a pandemic hit the industry. Projects have even tighter budgets and shorter time frames imposed upon them. What this means is that projects cannot afford delays, mistakes, or missteps. Every party involved needs to perform its work as efficiently and expeditiously as possible. Estimates must be accurate, augmented management staff must be applied to projects, and management must be ever vigilant to identify and resolve problems as soon as practicable. In today's construction environment, with its heightened demands, a continuous and concerted focus on productivity is crucial.

Whether the reader is a construction professional, a contractor retained to construct a building or renovate an existing one, or an owner with a business plan for its new building, the concept of productivity in construction is intertwined in so many aspects of the work. Here are just a few examples:

- A contractors' bid is based on certain assumptions/predictions/estimates concerning construction productivity.
- The assignment of time or durations to specific construction activities is based on assumptions/predictions/estimates concerning construction productivity.

Ted Trauner, Chris Kay and Brian Furniss. Productivity in Construction Projects, (1–8) © 2022 Scrivener Publishing LLC

- The type of equipment utilized for specific construction tasks is based on the productivity that must be achieved.
- The size and makeup of construction crews is driven by the productivity desired for a specific activity in a specific environment.
- The cost of the work flows directly from the productivity of the crews and equipment on the project.
- The overall project schedule is dependent on the productivity for each of the activities in the schedule.
- The overall profitability of the project is a direct function of the productivity that can be achieved on the project.
- The overall long-term viability and functionality of the project is a direct function of whether the contractor took "short cuts" in the construction process when it was discovered that the project's productivity, and thus the contractor's profits, were being adversely affected.

These examples demonstrate the pervasive nature of productivity within the construction project, as well as its daily importance. Unfortunately, the authors have seen too many parties in the construction project begin to truly focus on productivity only very late in the process—typically when they realize productivity has not been achieved, or not to the levels originally anticipated. Owners, contractors, sub-contractors, and construction professionals seem to express concern only when they finally realize there is a cost overrun in a specific area of work, or when activities are taking longer to perform than the time assigned. They become rightfully alarmed when forced to use longer work weeks and overtime on a daily basis. But by then, it is often too late.

In our collective experiences, we have encountered construction personnel who accept the general proposition that construction productivity is important on a prospective or forward-looking basis, but very few actually put that proposition into practice. Instead, their attention to productivity typically occurs when they look at it on a retrospective basis. By then, it may be too late to undo what has occurred. However, we can analyze what has occurred to determine the cause of the problem and, hopefully, seek a solution.

Perhaps the single most important lesson the reader should learn from this book is that any improvement in productivity, and the ability to adequately measure any shortcomings of actual productivity on the project, must begin with a comprehensive and detailed analysis of productivity on a prospective basis. We have seen time and again that owners, contractors, litigators, and consultants will expend hours and hours retrospectively attempting to quantify the amount of productivity that allegedly has been "lost" as a result of changes to the project during construction. If even half (or less) of those same hours were spent prospectively, the problem may never have occurred, and the additional legal costs (and time associated with a legal dispute) would have been avoided. Even if the problem persisted, the work to quantify any such losses is already more than half completed. We encourage the reader to adopt the more proactive approaches described in this book to avoid these kind of problems—and subsequent legal disputes—that have plagued so many construction projects in the past.

Performing the work of identifying and monitoring productivity analytics on the front end will increase the reader's chances of achieving a desirable result on the back end. Interestingly, what we describe as necessary work at the start of the project can actually reduce the amount of work that is necessary to be performed later in the project, or in subsequent litigation. Once an organization uses these processes at the beginning of the project, its employees will soon discover that the implementation and mastery of the productivity methods gets easier over time. Each project further increases an organization's body of knowledge and experience. If that organization then employs those lessons learned from preceding projects to its future projects, the organization will complete more projects on time and on budget, and earn greater profits—and future business. That is the essence of successfully harnessing productivity.

We recognize that some problems will nevertheless arise on the job site that result in the parties suing each other, either in court or in arbitration. Having spent considerable time on both, we also write this book to help the reader better understand how the dispute resolution process works, as well as the benefits and weaknesses of pursuing litigation and arbitration.

Thus, the second most important lesson to be learned from this book is that when parties find themselves in litigation or arbitration, they need to retain knowledgeable and experienced expert witnesses that employ credible and reliable methods of quantifying productivity losses. We have long believed that litigants in construction disputes made major mistakes when they did not attempt to use the "measured mile" approach to determine if there was a loss of labor productivity on a job site. The measured mile approach is predicated on the use and comparison of the actual facts contained within the data generated at the job site.

We do not want to see the next generation of owners, contractors and sub-contractors fall into the same trap as their predecessors, whose experts were apparently unable or unwilling to utilize a measured mile approach. As the reader will see, over the last 20 years the courts have essentially agreed with our point of view. We will show the reader the measured mile approach, provide actual case studies demonstrating its use, and how regression analysis may fit into that analysis process.

Along the same lines, we have long believed that litigants in construction disputes made egregious mistakes when they retained experts who relied heavily upon "industry standards," a collection of general studies authored by various groups over the last several decades. Here again, we do not want the next generation of owners, contractors, and subcontractors to make the same mistakes. As the reader will see, over the last 20 years the courts have agreed with our point of view in almost all instances. Many cases have been lost because the party's expert relied solely or significantly on such industry studies, without truly understanding the basis of those studies, and whether they were applicable to the subject project.

The third overarching message we emphasize in this book is the benefit of changing the workplace culture. Over the course of our respective careers, the authors have encouraged parties to collaborate to solve problems when they arise, rather than casting blame towards the other parties and engaging in defensive measures that result in lengthy and costly litigation or arbitration. The fact that the authors are owners, attorneys, and engineers collaborating on this book underscores our belief in the need for the kind of give-and-take, by everyone involved, to effectively and efficiently solve the problems that frequently arise on a construction jobsite.

Using that collaborative approach, we believe it is time for a different on-the-job culture, one that is significantly assisted by the implementation of new software and technology. With the adoption of a new culture and new software, parties can collectively create a schedule predicated on the same kind of units used in creating the measured mile approach. Such a construction project schedule would cause the parties to look at work per man units, and thus enhanced labor efficiency, at the outset of construction. Such a schedule would also lend itself to having the parties periodically review on-the-job productivity in real time and make real-time adjustments.

The basis of our recommendations come from decades of experience engaged in all facets of the construction process. Ted Trauner is a nationally recognized expert in the areas of scheduling, construction management, cost overruns and damages, construction means and methods, and delay and inefficiency analysis.

Ted has either managed construction or evaluated problems on virtually every type of project including transportation, water and wastewater treatment, process and manufacturing, power, medical, educational, commercial, correctional, hotels, condominiums, residential housing, and athletic facilities. For over 45 years, Ted has participated in the analysis and resolution of construction claims, managed many types of construction projects, and provided scheduling and training to the industry.

He has testified as an expert witness on delays, inefficiency, disruption, differing site conditions, excessive changes, extra work, termination, productivity, structural analysis, construction means and methods, and cost overruns and damages.

Ted lectures, conducts seminars, and has developed and presented training programs on multiple topics including construction claims, specification writing, partnering, and construction management to thousands of construction professionals throughout the world.

Ted is the author of the following highly-regarded construction texts; *Construction Delays, Third Edition; Managing the Construction Project; Construction Estimates from Take-Off to Bid; Construction on Contaminated Sites; Construction Delays (1st and 2nd Editions);* and *Bidding and Managing Government Construction.*

Prior to his work in the private sector, Ted was an officer in the U.S. Army Corps of Engineers for 11 years. He was Military Assistant to Construction Operations of the Philadelphia District, where he was Resident Engineer for major highway bridge rehabilitation, was involved with the construction of major gravity earth dams, and also advised on regulatory affairs for the District waterways.

Chris Kay was a trial lawyer for over 23 years and handled a wide range of construction disputes. As a construction litigator, he has extensive experience identifying the data and expert testimony needed to succeed in cases of all sizes and complexity. He also appreciates that a "win" needs to take into account the amount of time the client devotes to the litigation rather than to its normal business, and the cost of legal and expert fees. Thereafter, Chris became the first General Counsel in the history of Toys

'R' Us, and later its Chief Operating Officer. He was responsible for the construction of hundreds of new stores and renovation projects for existing stores, for the entire portfolio for Toys 'R' Us, Babies 'R' Us and Kids 'R' Us properties across the country. Chris later served as Chief Executive Officer for the New York Racing Association and spearheaded a number of capital improvement projects at that organization's three racetracks. As the owner of retail stores, malls, and sporting and entertainment venues, Chris knows that the corporate owner seeks to have the project completed on time and on budget, but not at the cost of the contractor taking detrimental "shortcuts" that may adversely affect the short-term operation or long-term viability of the completed facilities.

Brian Furniss has served in the construction industry for over 20 years, analyzing complex construction projects and providing expert testimony on scheduling, delays, productivity losses, and damages. As someone who has worked various construction sectors in projects ranging up to $15.5B, Brian has seen both exceptional management strategies and a fair share of missed opportunities. He is a licensed Professional Engineer (Industrial) in Florida, Texas, California, North Carolina, and Colorado, and a Planning and Scheduling Professional (PSP), Certified Cost Professional (CCP), and Certified Forensic Claims Consultant (CFCC) with AACE International. Brian is co-author of the book *Construction Delays: Understanding Them Clearly, Analyzing Them Correctly*, and has authored a number of articles for construction industry publications. He shares the experiences in this book knowing that every project provides a learning opportunity, and to help construction professionals increase the probability of project and company financial success.

We know of no other book where owner, attorney, and expert have collaborated together to explain and assess construction productivity, or inefficiencies at the job site. We have written this book primarily for the owner, contractor, and subcontractor—in that we write in practical yet specific terms. We also write extensively about what is required under the applicable law in construction disputes, so that the parties, their attorneys, and their expert witnesses can understand and appreciate those applicable legal standards as they begin to evaluate all the facts of their dispute—before they spend a ton of time and money in court or in arbitration.

Based upon our collective experiences, we offer the reader very practical insights and advice in straight-forward prose—starting with a chapter devoted to explaining what productivity is, how to measure it, and why it is important. We then devote a chapter to a simple but accurate explanation of a complex topic: what is a regression analysis and why it is important. We describe a regression analysis in terms we hope the reader will quickly understand, rather than relying on overly detailed explanations.

Precisely because there are too many post-construction disputes, we provide an explanation of how those disputes are resolved, and the best way for the reader to reduce or eliminate significant costs in that process. We also share with the reader the applicable law in construction cases, but not by providing lengthy string citations of several different court cases that force the reader to find and read the cases. Instead, we provide relevant facts of the cases we selected—cases we specifically selected for their informative or illustrative nature.

Finally, we provide guidance on how to achieve your goals of "winning" your case—in settlement or in court. Employing our recommendations will likely lead to fewer lawsuits and arbitration claims after the job has been completed. Even if there are some unresolved problems that cannot be reconciled at the end of the construction process, we believe this approach will narrow the issues in dispute and make for a less costly resolution process. None of the authors, including the attorney, want needless or needlessly expensive litigation. We hope this book will help some readers avoid such unpleasant and costly situations.

2

Productivity in Construction

During much of the past two centuries, productivity was a driving force in this country and throughout the world. The ability to produce and distribute goods in an efficient and cost-effective manner drove the financial success of nations. The United States became a world power not solely because of military might. Its rise was also driven by the successes it had in industry. American industry moved from hand tools to assembly lines to robotics in search of higher productivity. The U.S. applied this same production focus to agriculture, not only feeding this country, but also being able to export food to the rest of the world. The ability to produce efficiently, to achieve high levels of productivity, has been key to America's economic success.

Workers have experienced tremendous changes during the past century. The basic manner in which they earn their livelihoods has been dramatically altered from farming and hunting in the 17th century, through the factories of the industrial revolution in the 18th and 19th centuries, to the Internet workplace of the 20th century and today. Many of us remember our parents putting in their nine to five in a routine that varied little. But today, our work environment is drastically different. We rely on computers and robotics to handle many of the chores we formerly labored through. The United States, in particular, has moved from an industrial base to a service-oriented economy, but the same focus on productivity applies—and has been the key to many companies' success.

The Continuing Importance of Productivity in Construction

Despite operating in the Internet Age, we still rely on manual labor to construct our homes, high-rises, power plants, airports, and highways. In fact, construction is one of the few career paths that has not changed substantially

Ted Trauner, Chris Kay and Brian Furniss. Productivity in Construction Projects, (9–18) © 2022 Scrivener Publishing LLC

in the last century. Though the industry benefits from more sophisticated equipment, more advanced designs, and a greater choice of materials, it still relies on construction workers to put the pieces together to create a functioning system to meet our needs. But the same focus on productivity that now applies to both manufacturing and service-oriented businesses can—and should—apply to construction project planning and execution as well.

Construction, like any other business, is driven by the bottom line. Construction managers, general contractors, subcontractors, and specialty contractors are in the construction business to make a profit. One of the most significant factors affecting that profit is the productivity or efficiency of the construction tradesmen working on the project and the construction equipment utilized on the project. Consequently, construction productivity is extremely important to contractors.

Productivity is also important to owners of construction projects. If workers on a project perform in an inefficient manner, the work may be completed later than required by the contract, much to the disappointment and lost revenues of the owner. In such situations, owners may also face claims for additional payments advanced by the contractor for lost productivity. Owners also need to be concerned about situations where the contractor takes short cuts to expedite the project. These shortcuts, in many cases, diminish the useful life of the building or compromise some aspect of the facility's operation—all of which adversely affect the owner's investment in the property. Architects who design facilities that cannot be constructed efficiently may drive the cost of the project so high that it will be cancelled for budgetary reasons. Hence, all parties involved in a construction project have an interest in maximizing productivity on the job.

What is Productivity?

In its simplest form, productivity or efficiency is the relationship between a given outcome and the resources expended to produce that outcome. In simpler terms, it is units of work over units of input (or the inverse). For example, presuming a contractor is diligently tracking work on a project, a daily record of work performed may show that 400 feet of 4″ pipe was installed with 40 man-hours of effort. Simplifying that means that the work crew was achieving a measured productivity of 10 feet of pipe per man-hour (400 feet divided by 40 man-hours). The inverse of that would be 40 man-hours divided by 400 feet of pipe or one tenth of a man-hour per foot of pipe

(40 man-hours divided by 400 feet of pipe). We can readily relate to productivity in manufacturing when we think of how we might make a certain number of widgets during a certain period. Because the manufacturing facility requires a certain amount of resources for each minute that it operates, we would want to produce as many widgets as possible for every minute we are running our production line. Due to the fact that assembly line tasks are generally repetitive, with few (if any) variables, the efficiency of the manufacturing operation can be measured and evaluated easily.

By now, it should be apparent that productivity is a ratioed comparison of quantity of work achieved to hours expended to complete that quantity. We have seen the terms "productivity" and "production" used interchangeably. While the output measure is typically the same between the two terms (quantity of work complete), the input of the two terms is very different. As a result, the production is not the same as productivity. Let's expand on this for a minute.

Production is primarily focused on output. It is often expressed as the production of an overall steel fabrication plant by month, a country's output for the year, or the amount of rebar placed in a day. Production is output compared to a duration-based input. The production rate is the output expressed against that unit of time. The output of a cell phone manufacturing plant may be 300 phones per shift.

What production does not explain is the resources input to achieve that outcome. This is where productivity comes into play. Productivity changes the input from a duration-based unit to a resource-based unit. While the production rate may remain constant when comparing two time periods, the productivity rate may vary greatly. To illustrate this point, let's presume that in two different time periods, the production rate of that cell phone plant remained consistent at 300 phones per shift. However, the productivity during the two shifts was different. In the earlier time period, a single production line was used that had 5 workers and each worker worked 8 hours per shift (let's ignore the use of robotics in phone production for the sake of simplicity). The earlier time period had a productivity of 300 phones per 40 labor hours, or 7.5 phones per labor hour (300 phones divided by 40 hours). In the later period, two production lines were used as the plant was anticipating shutdowns for preventive maintenance and the plant wanted to maintain its shift production rate. Each production line had the same hours per shift. So in the later period, 300 phones were produced using 80 labor hours (two shifts at 40 hours per shift), resulting in a productivity of 3.75 phones per labor hour (300 phones divided by 80 hours). Despite production being the same between the two shifts, the productivity in the earlier shift was twice as good, or 100% better, than the productivity during the later shift. Conversely, one may argue that the preventive maintenance caused a productivity loss, or inefficiency, of 50% in the later period. This example demonstrates the differences between production and productivity rates, and that even when there are equal production rates there may be very different productivity rates.

Productivity in construction is a bit more difficult to define because, unlike manufacturing, construction tasks that may appear repetitive, actually may vary to some degree. For example, a worker on an assembly line normally is repeating the identical same task with each widget that passes his/her station. A construction worker, on the other hand, may be repeating similar work, but a variety of factors will make each task a bit different and, therefore, not executable in precisely the same manner every time. These variations can be significant. For example, an electrician may spend an entire day installing 1″ conduit. It may seem that because the conduit is all the same size, the productivity should be uniform throughout the entire day. However, the conduit runs are likely to be located in different areas and at different heights. Some conduit runs may be waist high in an open area. Other conduit runs may be ten feet in the air, requiring the use of a ladder or scaffold to perform the work. Still other conduit runs may be vertical, while others are horizontal. And each conduit run may be anchored or supported in a different fashion. There may be hundreds of variables that enter into the manner in which the conduit is installed: ten-foot sections or smaller sections; numerous elbows; working around ductwork or plumbing lines; no drywall installed, or drywall installed on one side; the presence of other workers in the area; hot or cold working conditions; damp or dry working environment; well-lit or dark workspace. The list of real-life variables at construction sites goes on and on. Each of these factors can affect the speed or efficiency that the electrician can achieve in installing the conduit.

In the interest of further illustrating some of the variables that distinguish construction productivity from productivity in the factory setting, another example is worthwhile. Suppose a construction contractor is excavating soil from an area. The quantity of soil that can be excavated during any unit of time is dependent on many different elements. The soil may vary in hardness or degree of compaction. Some areas within the excavation may be wet, while others are dry. The operation may be in an open area for some portion of the work and far more restricted in other areas of the excavation. As a consequence, the speed at which the material can be excavated may vary over the course of the day.

The problem in analyzing construction productivity is that, in contrast to the manufacturing facility, the work cannot be isolated and maintained in a controlled environment. This does not mean that construction productivity cannot be managed or controlled. In fact, many of the most significant factors that affect construction productivity are within the control of the project management team. It is precisely these factors—factors that can and should be controlled by the project management team—that are addressed in this book.

We recognize there are some areas over which we have no control, like weather. But even in those areas, we can properly estimate the weather's potential effect on the workers' productivity during inclement weather, and then plan accordingly. The key to maximizing productivity, in both those areas over which we have control and those where we do not, is to properly plan ahead of the project's commencement of construction and continually evaluate that plan throughout the project. Let's take a moment and identify factors that are controllable and non-controllable.

Controllable Factors that Affect Productivity

There are many factors that we can control that will affect the productivity achieved on the project. While these are discussed in more detail in the remainder of the text, it is worthwhile to briefly identify them here.

Construction Schedule: The schedule we create for the project dictates the sequence of work and the time allotted to complete specific activities. Obviously, both the sequence and the duration allowed can affect the productivity achieved. If we structure the work such that our sequence of operations is not optimally efficient, we will see a corresponding reduction in the productivity that can be attained. For example, if the project manager plans to drive underwater piles for a pier and excavate or dredge the material from between the piles after they are driven, this may be less efficient than first dredging the material and then driving the piles. While there may be good reasons for choosing this sequence, the project manager must accept the reduced level of productivity. Consequently, our planning and sequencing of the work directly affects the productivity we will achieve.

Equipment: The type and amount of resources we assign to specific tasks will affect the productivity of the task. If the project manager utilizes equipment that has limited production when it is possible to use more efficient or more productive equipment, the direct consequences must be recognized and accepted. For example, if a contractor chooses to perform excavation with a small backhoe when a larger piece of equipment would yield better rates of production, that should be factored into the plan. Problems arise when an estimate is based on optimal production with equipment that is never utilized. Similarly, if the crews assigned are either too large or too small, the efficiency of the operation can be adversely affected. Resources are an important element of the productivity equation.

Application of Resources: The duration allowed for a task and the number of shifts to which resources are applied will also bear on the efficiency of the work. We address the use of overtime, extended workweeks, and multiple shifts in more detail later in the text. Suffice it to say that there has been general acceptance that excessive overtime, extended workweeks, and multiple shift work may reduce the productivity of the workers. In most cases, we can control the use of these approaches.

Site Layout and Management: The manner in which we organize the site also can affect our efficiency. This is different than our organization or sequencing of the work itself. In the construction industry, it has been demonstrated that how the site is organized, such as where we store materials, where we marshal deliveries, how we distribute materials, and how we make tools and supplies available, can impact the efficiency with which the work is performed. For example, on a large process plant project, a contractor might set up its parts trailer on a remote corner of the site in order to allow easy access for deliveries. Unfortunately, that contractor may later discover that as the workers require parts for construction, the remote location may require them to walk a long distance to acquire these parts. This might result in a reduced productivity due to the time spent acquiring the necessary components for the construction.

Labor Supervision: The amount and the quality of supervision employed on a project can also have an impact on the efficiency achieved. More experienced foremen and superintendents can often recognize better and more efficient work methods for their crews to utilize.

Non-Controllable Factors that Affect Productivity

Weather: Almost all will agree that cold, snow, extreme heat, high winds, rain, and other climate conditions can alter the productivity of construction workers who are exposed to them. While there are some measures we can take to mitigate the impact of weather, we cannot eliminate all impacts. We may "tent" or enclose work areas to reduce the exposure of workers from the weather. We may incorporate heaters in the work areas to raise the working temperatures to more acceptable levels. We may create additional drainage on the worksite to reduce the amount of water infiltration into the work area. We may take different steps to lessen the impact of weather, but we cannot completely eliminate the negative effects of weather on our work, and we cannot completely control the occurrence of unfavorable weather at the worksite.

The fact that adverse weather conditions reduce the level of productivity of construction workers does not necessarily result in a reduction in expected performance. Because certain weather conditions are known to exist in specific areas year after year, a contractor can estimate and plan the work considering the likely weather conditions. For example, if we know that our project site typically experiences ten days of snow as a yearly average, we can estimate and plan for that level of interference to the work. In other words, our expected level of productivity should be based on the anticipated conditions of the job. While we cannot control the weather, we can manage the risk associated with it by estimating and planning for what should be expected.

This example highlights one of the major differences between productivity in construction and productivity in manufacturing. In a factory, we can control many of the environmental factors that affect productivity, but on a construction site we do not have that same level of control. Also, we must recognize that the occurrence of inclement weather that negatively affects our performance does not mean that we experience a loss of productivity. A loss of productivity would occur only if the weather differed from what was normally anticipated for that time of year, and thus not made a part of the initial schedule for the project.

Skilled Workforce: Another factor we cannot control is the availability and skill level of the workforce. It is not uncommon for projects to face shortages of qualified workers. This is particularly true on very large projects or in areas where the overall level of construction activity is high. Many areas of the country have had "boom times" where both housing and commercial construction have exhausted the available workforce requiring "travelers" to offset the shortfall. Better management and planning can mitigate problems with shortages of qualified workers. However, we cannot overcome all the shortcomings of this problem by management alone. As with weather, a loss of productivity because of the workforce available is most meaningful when it is based on a comparison of what should reasonably have been anticipated when the project was bid on. If the construction contractor is aware at bid time that various trades are in high demand with limited availability, the plan of execution and the bid should reflect this. Planned levels of productivity should account for any variations in the anticipated availability of trades.

Worksite: Construction productivity may also be influenced by the available access to the worksite. For example, one excavation project may have available ingress and egress from virtually all directions. This allows the contractor to exercise great flexibility in how it executes the actual excavation operation. Alternatively, another excavation project may have limited access. It may require the contractor to enter the site from one point, excavate material, and exit the site through the same route. Obviously, the daily productivity of the equipment and operators will vary significantly between these two projects. That does not mean that there is a loss of productivity. It merely means that the productivity that can be achieved differs because of the physical parameters of the project site. While we may not be able to have optimal access to every site, we can and should plan for a realistic, achievable level of productivity based on the known site conditions.

Design Complexity: The complexity of the design and construction of a project can also affect the level of productivity. Electrical work on a process plant may be much more difficult than in a warehouse facility because of the complexity of the design or because of the physical configuration of the electrical installation compared to electrical work. Once again, the manager can plan for this to ensure that its estimated productivity is achievable. This does not mean that the electrical work on the more complex project exhibits a reduced or loss of productivity. Rather, it highlights that productivity in construction varies from project to project depending on the specific characteristics of the job. Productivity should be compared against a baseline productivity that considers the characteristics known at the time of bidding, not against other projects with different characteristics that might allow for different levels of productivity.

The Focus on Productivity Should Be Continuous and Based on Hard Data

Productivity in construction is often addressed in three distinct contexts: the planning stage, during the execution of the work, and at the project postmortem. Unfortunately, the priority given to these by project managers is commonly the reverse of what should be employed. Often, the least amount of thought and effort is given during the planning stage. Slightly more attention is paid to efficiency during the work execution stage. And the most emphasis on productivity is applied during the project postmortem, usually in a situation where a construction claim has arisen. As a consequence of this questionable if not flawed approach, much of the literature on the subject of construction productivity has been written in the context of claims and disputes.

In reviewing the literature on construction productivity, numerous "laundry lists" are cited that purport to identify the factors that cause or affect the loss of efficiency experienced on a project. Many of these factors are discussed and evaluated only when claims are asserted, rather than at the time the alleged problem occurred. Some of the factors that are most often presented in construction litigation include the following:

Fatigue
Morale and attitude
Stacking of trades
Reassignment of manpower
Crew size
Dilution of supervision
Learning curve
Errors and omissions
Beneficial or joint occupancy
Logistics
Delays
Shift work
Cumulative effect of multiple changes
Overtime
Compression

Items from the preceding list are presented to explain an adverse effect on productivity that resulted in a corresponding financial loss. Often, little, or no "real-time" quantitative analysis is performed or presented to support the factors alleged or the inefficiency claimed.

In the succeeding chapters of this book, we address the subject of construction productivity as it relates to planning and executing construction tasks before and most importantly during a project, as well as measuring and evaluating changes to productivity during the postmortem. There are many steps a manager can take, and an owner should demand, of the construction manager in the planning and execution stages of a project to better control the productivity that the project experiences. This requires an understanding of the factors that affect productivity as well as a conscious and continual commitment to focus on ways that the process can be managed to improve efficiency from the planning stage through to the completion of the project. It also requires some understanding of the means to best evaluate productivity while working on the job, and the data and technological tools currently available to assist in the productivity evaluation process.

The rest of the book provides methods and examples to measure and calculate productivity losses. Other methods, including industry publications and studies referencing productivity loss, are also discussed. However, there are no sections on the modified total cost or total cost methods. This is not a mistake, and the authors are not ignoring these methods. While courts and various jurisdictions have accepted these methods as ways to measure the additional costs resulting from productivity losses, they are not methods of measuring or quantifying productivity losses. The increased technological and software advancements have made the tracking of productivity metrics more commonplace and, we believe, will further reduce the need to use these methods. As such, these cost-based measurements will not be discussed throughout the rest of this book despite their acceptance in certain courts and jurisdictions.

In assessing productivity on a construction project, nothing is absolute. All of us have heard "absolute statements" related to construction and to construction productivity. For example, "Everyone knows that overtime causes a reduction in worker efficiency." Or "Everyone knows that encountering a differing site condition will reduce the efficiency of an operation." But these so-called "absolute statements" may be totally false assumptions. Until one can test, analyze, and assess the productivity demonstrated, such so-called "absolute" assumptions should not be made. While many factors can affect work and can be shown to have affected work on some projects, that does not mean that those same factors will have even a similar effect on any other given project. As a result, both the prudent owner and the professional manager will rely on real-time data for their project to assess productivity and promptly make whatever changes are deemed necessary to achieve the best possible result.

3
Measuring Productivity

A meaningful discussion on how to improve productivity starts with an understanding of the nature and level of our existing productivity. Hence, we must determine this before we can make improvements. Similarly, we cannot make the assertion that a change or some other factor adversely affected the productivity of a trade on a project unless we have an objective measure. Therefore, we have some work to do. Let's look at the major participants on a construction project and objectively assess how productivity can be determined. For simplicity's sake we will take the two major participants: the owner and the contractor. Within each of those, we will make some subdivisions in order to have a clearer picture.

The Owner

How does an owner track productivity on a construction project? It depends on the size and nature of the business of the owner, and the resources the owner is willing to commit to the project. What are those resources? An owner that performs numerous construction projects, such as a governmental agency (including for example, the U.S. Army Corps of Engineers, the Office of Veterans Affairs, the City of New York, the various Federal and State Departments of Transportation and/or State Highway Administrations), normally has an existing staff of resources such as construction inspectors, resident engineers, design staff, and quality control personnel. Private owners that build a number of facilities, such as big box retailers, mall developers and office building developers, generally have both real estate and construction departments within their organizations. These departments will have employees that not only select and contract with contractors and architects, but they monitor the construction process

Ted Trauner, Chris Kay and Brian Furniss. Productivity in Construction Projects, (19–32) © 2022 Scrivener Publishing LLC

and progress. Depending on the nature and location of the project, they will also hire the same kind of third parties noted previously.

Owners that do not construct buildings as an integral part of their business (and therefore perform fewer projects) normally enlist the services of supporting third party organizations such as owner's representatives, project management companies, agency construction managers, or oversite by architectural firms. There is often confusion when assessing the similarities and differences among these different types of organizations, all of whom claim to provide necessary and important assistance to owners in the construction process.

Owner's Representatives/Project Management Companies: To some in the business, these terms are interchangeable, which it is why it is important to understand the actual work they perform for owners. Owner's representatives are usually charged with monitoring—but not managing—(1) the design development process (serving as the primary advisor to the owner for architectural issues, rather than the architect); (2) the project budget; (3) the bidding process; (4) the schedule (in some but not all companies); (5) construction pay applications; and (6) project close out and occupancy.

Agency Construction Managers: Quite a few years ago, the term "Construction Manager" came into vogue. There were many reasons for this, but the initial concept was to provide a construction company or someone to work with the owner starting at the design process and continuing thru the entire construction. In this manner, the construction manager (CM) would be an available resource to the owner to provide preliminary estimates on a design and to suggest cost savings that could be implemented during the design process. The concept envisioned a partnership between the owner and a construction contractor (the CM) from the beginning of a project instead of the more typical design-bid-build approach. It should be noted that it was common practice in the design-bid-build process to require the General Contractor to perform a specified percentage of the project work with its own staff (commonly at the 10% to 15% level). This requirement went by the wayside in the CM process. This then allowed the CM to subcontract all work on the project. As the CM process evolved, Agency CMs entered the scene. An Agency CM is a company that works with the owner from the project inception through the entire design and construction process. The major difference is that the Agency CM (ACM) does not subcontract any of the work but provides more oversight.

The ACM concept evolved from the belief that an ACM would be less biased since it was not directly involved in the subcontracting of work and therefore, would not be "protective" of any subcontractor bids or requests for additional compensation. We will not offer any opinion on the best method for an owner to manage a project other than to note that the method of compensation for the ACM is a major factor to consider before going that route. A brief explanation is worthwhile.

The ACMs may have their compensation structured in various manners. For example, the ACM nay charge on an hourly, weekly, monthly, or use basis at predetermined rates. In simple terms, if the ACM requires two people on the project site, the ACM bills for the time the two representatives work. In one project that the author consulted on, the ACM was compensated based on a percentage of the construction costs. Reflect on that for a moment. If the ACM is paid on a percentage of the total construction costs and the contractor requests a change order that will increase the cost of the contract, the ACM will receive higher compensation. This belies the "loyalty" to the owner and may not serve the owner's best interests. The bottom line is that if an owner decides to use an ACM, the compensation must be structured to support the goals desired by the ACM approach.

Architectural Firms: Many architectural firms will propose services for construction oversight. This usually includes tasks such as: periodic site visits and inspections, review, and approval of pay requests, review and approval of submissions, review and approval of change order requests, responses to requests for information (RFIs), and project closeout. Compensation for these types of services is usually on an hourly basis at predetermined rates for the level of staff. The inherent risk in the owner's use of the architectural design firm to provide construction services is that the designer may have been motivated to "protect its design." For example, if a contractor submits a change order request that may have been occasioned by a design error, the designer might not be willing to approve it lest it reflect poorly on the initial design.

Whatever approach the owner chooses to represent its interests on the project, it must choose a firm that has experience and satisfied clients. All too often, the owner performing fewer projects decides to go cheap—hire a "clerk of the works" and engage them to make only periodic site visits. Going cheap may be far more costly in the long run. No amount of exculpatory contract language will make up for consistent and careful oversight of the actual project.

Let's start with the constant construction owner such as a Department of Transportation. Let us further accept that the owner has full time inspectors on the project. If any of our readers have read inspection reports from a project, they will not be surprised when we state that most of those inspection reports can be disappointingly substandard. All too often the inspection report includes the date, day of the week, a brief note about the weather (temperature, precipitation, etc.), a very general description of the ongoing work, a listing of the contractor and subcontractor personnel on site (usually by number of people), and a note on visits to the site by other personnel. No meaningful information on the daily or weekly progress of the work vis-à-vis the schedule, or problems/delays encountered, is documented. After reading one of these typical inspection reports, it makes one wonder what the inspector did for the remaining 7½ hours of an 8-hour day.

But, regularly collecting contemporaneous and comprehensive relevant data—with precision—is critical in determining accurate levels of productivity in real time. That determination can help the owner and the contractor together overcome construction delays as they occur to the mutual benefit of both parties. Alternatively, that data collection process can help the owner and contractor when claims are filed against each other in post-construction litigation or arbitration.

For example, years ago, a State Highway Administration (SHA) was constructing a major bridge project. The bridge crossed a waterway and had 21 concrete piers for the substructure. The concrete piers had precast concrete piles, pile caps, etc. The project experienced a significant delay and led to litigation between the general contractor and the SHA. The major claim by the contractor was that cobbles and boulders were encountered in the subsurface conditions while driving the piles. This differing site condition caused major delays and a significant cost overrun because of reduced productivity in the pile driving.

Let's give you some more pertinent facts.

First, the original subsurface investigation of the site had numerous subsurface borings. On 7 of the 21 piers, the borings clearly showed the presence of cobbles and boulders.

Second, the borings from the subsurface investigation were not included as part of the contract bid documents. Therefore, the bidders were not privy to that information and no bidders had asked the SHA if any such information existed.

About this point, the reader is probably thinking that the claim by the contractor sounds pretty darn good. But, as always, the devil is in the details.

The SHA had more than one inspector on the project. One inspector was solely responsible for reviewing the ongoing pile driving operations. This inspector was unusual. He kept very precise and detailed records on the pile driving. His inspection reports specified the exact pier and pile being driven. He chronicled the exact time spent driving each and every pile. He recorded the exact time expended to move the pile driver, to change the cushion block, etc. Simply put, he recorded every minute spent on pile driving and exactly how it was spent.

The information recorded by the inspector was FANTASTIC! Based on this information, a detailed analysis was performed to determine whether the seven piers with cobbles and boulders took more or less time or increased or decreased productivity more than the other fourteen piers that had no cobbles and boulders. In the analysis, all factors were considered, including the efficiency of the pile driver, the crew size, the length of the piles, and other relevant data.

The results were illuminating—and dispositive. Based on the detailed inspection reports, the piles in all seven piers were driven as productively or better than the remaining fourteen piers. Simply put, the presence of cobbles and boulders caused no reduction in the productivity of the pile driving operation. As a direct result of the inspector's diligence in recording all pertinent information, the majority of the contractor's claim was found not to have merit and was denied after a trial that lasted nearly three months.

The moral of the previous story is as clear as it is crucially important. The owner needs to have an adequate inspection presence and those inspectors need to perform well above the norm we have historically experienced, to a higher standard that must be set by the owner or its representative. Bear in mind, that inspection effort can be conducted by direct employees of the owner or the owner's representatives, an agency CM, a project manager, or staff from the construction services with the design firm.

No matter who collects the inspection data, decisions should be made at or before the initiation of the project as to the specific type, scope, and depth of information that should be maintained, as well as the frequency of that project monitoring and documentation. Once begun, management must review the reports at predetermined intervals to ensure that information is properly and completely recorded in the detail required.

In the name of clarity, and to make life a bit easier for all parties concerned, we suggest that the owner and/or the owner's representative prepare a daily report form that lists all of the information that the owner may require and that should be maintained. But don't stop there. Also prepare a sample of an acceptable report filled out and a sample of an unacceptable report. Sorry, your job isn't over yet. Now the owner and/or the owner's representative must review the daily reports to ensure the information is correctly recorded and the acceptable level of detail is provided. Once all parties are performing correctly and collaboratively, it is essential that the owner and/or the owner's representative continue to review the progress reports and "counseling" any parties when their reports are inadequate.

One of the benefits of operating in the internet age is that there are many electronic apps and software that allow us to record information easily, in real time, and in reasonable detail. The desired information can—and should—be captured digitally (include video, especially when documenting a possible problem that could lead to a loss of productivity). In fact, it would be wise for whomever is managing the inspection staff to establish the detail required in reports for each portion of construction, and then train those responsible for using the technology to record the information properly and completely. For example, if the electrical subcontractor is working, the information recorded cannot be limited to the man-hours expended per lineal foot of conduit; instead, the size of the conduit, the length of the runs, the height of the conduit, if scaffolding was required, and the number of bends/joints would all be meaningful information that should be part of the record.

In light of the fact that we have stressed the importance of collecting job-specific data, it would be disingenuous for us to now list every detail of information that should be tracked. But this book can stress that the project's manager must give careful thought to the information required and the units that should be used to allow a determination of productivity from that information.

The Contractor

For discussion purposes, let's presume our project has either a general contractor (GC) or a construction manager (CM). Realistically, there is little difference between the two. Historically, construction projects that retained general contractors required that a set percentage of the contract work had to be performed by the general contractor's staff whereas a construction manager had no requirement to perform a percentage of the work with his/her own staff. Either way, the GC/CM represent the entity that actually will be performing the work.

One would think that when the project is bid on by the GC/CM that productivity documented historically would be a key factor in the preparation of a bid. Let's not be naive. The majority of GC/CM bids will be based on a compilation of bids by subcontractors. In practice, the GC/CM will distribute the various parts of the contract work by trade discipline (electrical, HVAC, plumbing, steel, etc.) and request bids from the respective subcontractors. Estimating and bidding a portion of the work on a project can be performed in many different ways. In most cases the subcontractor will (or at least should) use its own historical data to arrive at a price for submission. The most significant question is "on what experience is the historical data based?" It may be based on units, dollars, or a number of factors. All too often, it is not based on accurately measured historical productivity—but it should be.

If not based on actual and accurate historical productivity data, what then, is the basis of most subcontractor's bids? It could be numerous and different answers depending on the subcontractor. The bid could be derived from estimating manuals (such as RS Means). It could be based on a man-hour estimate and quantities. It could be based on cost and quantities. We know from our experiences that there are various different approaches used by subcontractors when developing a bid.

Even at this early stage of this book, it should already be apparent that the accuracy of the bid is directly dependent on the historical data amassed by the subcontractor over a period of time. But is that enough? Is the subcontractor's bid predicated upon the nature, quality, and detail of its experiential data, and then customized to take into account the specific issues for this project? Allow us to elaborate a bit. Presume that a plumbing contractor is preparing a bid on a project. If the subcontractor's historical data shows that past projects have cost X dollars per foot of 4″ pipe, is that specific enough? Probably not. Is the historical data based on all 4″ pipe installed regardless of location? Or is the data further subdivided into 4″ pipe installed from a floor level or on a scaffold?

Then there are site-specific considerations that should be incorporated into the subcontractor's bid. Does the historical data account for where materials will be stored and distributed? Obviously, if the storage trailer for this project must be located several hundred feet from the installation, what dollars will account for moving the pipe from the "yard" to the installation point? Do different workmen display different installation rates? If so, what rate forms the basis of the historical data and what workman will be available for this project?

By now, the reader should be getting our message. The better our historical data, particularly when coupled with site specific facts for a given project, the better able we are to prepare an accurate estimate and the fewer problems we should have on the project. But the saga doesn't end yet. What foremen will be available? What project manager will the contractor be using to run the project for my specialty? How much time is allotted in the schedule for each respective activity and what number of workmen will be needed to accomplish that work in the allotted time? All of these factors can affect the productivity that we are able to achieve on the project, as the following story illustrates.

A power company in the western United States had undertaken a project to run new high-tension lines across mountainous terrain. The physical project required the use of towers that were a combination of freestanding towers and guyed towers. (For a freestanding tower, think of the Eiffel Tower – 4 legs on foundations with a steel lattice structure rising to the apex. For a guyed tower, think of a steel lattice "post" that is supported symmetrically on 4 sides by steel guy wires.) In order to expedite the project, the owner had an engineering firm complete the design and the owner then preordered the fabrication of the steel for all of the towers. The job for the successful contractor was to erect the towers, run the lines, appropriately sag the lines, and do the connections at both ends. In essence, since the steel towers were designed and fabricated, this became a large-scale "erector set" type of project. We noted previously that the project was in mountainous terrain. What this meant was that men and materials had to be helicoptered to each worksite. Then the fun began!

When the contractor performed the tower erection, he noted what he believed were significant errors in the fabrication of the steel. The lattice structure for each tower (both guyed and freestanding) required the bolting together of numerous pieces. In some locations, there were as many as five pieces bolted together at a single point. The contractor asserted that the bolt holes in the steel did not always line up and he was forced to re-drill or ream the holes such that all of the overlapping pieces aligned enough to allow bolting.

A claim was submitted by the contractor for several million dollars for this alleged extra work. In his claim, the contractor included a count of every misaligned bolt hole. (Note: if five pieces were at a bolt hole and a bolt could not be inserted through, then this was "counted" as five mis-fabrications.) From just a commonsense view, asserting that misalignments caused more work makes sense. But, as a wise man has said to us many times, "the devil is in the details." Having been retained by the owner to review the claim, it seemed that towers with a higher number of "mis-fabrications" should be the ones with the most man-hours expended. But, it is always wise to check the facts. When we checked out this theory, it wasn't supported by the facts. Towers with a very high number of "mis-fabrications" did not take more effort (man-hours) than towers with far fewer "mis-fabrications." We must require the same precision of ourselves to make an accurate assessment as we do for those working on the project, such that we segregated the guyed and freestanding towers so that the comparison was apples to apples.

Needless to say, this was befuddling. We ran a regression analysis (more about regression analysis later in this book). There was no correlation between the number of incorrect bolt holes and the man-hours expended to erect the towers. We asked ourselves "what factor was most realistically controlling the amount of effort on the towers if no correlation existed between the number of alleged incorrect bolt holes and the man-hour effort required to erect the towers?" Based on experience, we performed a variety of other regression analyses. These findings showed that one specific foreman always achieved good results (fewer man-hours) regardless of the number of misalignments and one other foreman always achieved the worst results (more man-hours) regardless of the number of misalignments. In the final analysis, the controlling factor for the majority of the excess man-hours was which foreman was heading up the specific tower erection.

Ah, yes, the devil is in the details, but in this instance, the result was devilishly dispositive for the contractor's ill-fated claim. In this particular instance, the contractor's claim was undermined solely by the relevant data collected daily at the project site.

But the same relationship between owner and contractor also applies to contractor and subcontractor, such that the same collection of relevant data can be used by a contractor to respond to the claims of a subcontractor.

Along the same lines, just as the owner is well served to work with the contractor on the development and subsequent monitoring of the schedule and related productivity, the contractor is equally well served to have the subcontractors participate in the scheduling and monitoring process.

It is clear that GCs/CMs rely on the accuracy of the bids received from the subcontractors when submitting their bids on the project. The respective subcontractors when bidding on the project have (or should have) developed a base of historical data for the various types of work they will be performing on the project. The quality of that historical data is of tremendous significance with respect to the accuracy of the bid. That data can be based on many different factors. It could be derived from a simple dollars per square foot, or cost per unit (such as feet of 2″ pipe), or crew hours per unit, or man-hours per unit, or many other approaches.

Also in the mix, is the fact that subcontractors bidding on a project must make certain assumptions during the preparation of their bid. Let's take an example just to highlight this area. A drywall subcontractor submitted a bid for the installation of drywall on a 10-story hotel project. One of the many assumptions inherent in this bid was that the subcontractor could work a full floor at a time and that access for that entire floor was available. During construction, a design error was discovered that necessitated a redesign of portions of the exterior facade and the interior layout of the floors.

As a consequence of this unfortunate result, the drywall subcontractor worked on partial areas of the floors away from the facade, and then had to return to each floor as the redesign work was completed. Access to an entire floor was never available, and most floors required three to four return trips to perform the work. Another consequence of the revised sequence of work was that some areas on each floor could only be partially completed until one of the later return trips. The end result was the submission of a claim by the drywall subcontractor to the CM, then to the owner and architect.

What do we take away from this? Most importantly, if a bid is based on certain assumptions, they should be stated in the bid submission. But perhaps even more importantly, some entity on the project should have been tracking and documenting the actual productivity of the drywall work both before and after the design change. If that had been performed, the task of calculating any differences in productivity and any consequent cost change could have been a relatively easy task—yet one more reason why all parties involved should periodically measure productivity in detail on the project.

Another factor in this equation relates to the project schedule. Most contracts from the GC/CM to the subcontractors include language that the subcontractor will provide all necessary resources to achieve the schedule that is prepared by the GC/CM. Well, sports fans, just saying so doesn't necessarily make something true. Ask yourself a question. "How often does a GC/CM prepare a project schedule with actual input from the subcontractors who will be performing the work?" NOT VERY OFTEN! Yet, the durations that are assigned to specific activities in the schedule must be based on the resources each subcontractor will be applying to each activity.

Let's keep one thing in mind—we are trying to avoid problems. Therefore, the GC/CM should invite, welcome, and insist upon input and participation from all of the subcontractors with respect to the schedule precisely because schedule durations drive the resources (manpower, hours per day, and equipment), and the resources affect the price and the cost of the output (dollars per unit).

Although we won't launch into a dissertation on project scheduling in this book, we will provide the following observations:

1. The majority of project schedules are prepared by a "scheduler" who will not be performing any of the project work.
2. It is relatively simple to make the schedule portray that the project is "on time" with any number of manipulations made by the participants attempting to mask any lack of productivity occurring on the project.
3. By the time project personnel realize that the project is not "on time," it is too late to correct the situation, or the correction is extremely expensive and moves the entire ballgame toward a court or arbitration. You don't want to have the last activity in the schedule to be "appear in court and testify."

Since we have stressed the importance of using the specific market conditions and job or site-specific challenges to define the best method to thereafter measure productivity, we are not now going to tell you the best way to track productivity on your particular project. That must be your decision based on the work you are performing. We can offer, however, some guidance, including the following advice we have previously offered:

1. Everyone (owner, owner's representative, contractor, and subcontractors) involved in creating the activity durations should base the schedule on the reasonably anticipated productivity for your specific situation.
2. The activity durations and sequence of work must be based on historical data, current market conditions, and site-specific issues and challenges.
3. Develop a comprehensive plan to monitor the work, with clear direction to all parties regarding the amount, scope, timeliness, and detail required, as well as the times when the data will be collected in the monitoring reports.
4. Use technology to your advantage. The present era of digital phones and tablets should make it far simpler to record whatever type of data you need to measure your productivity. The parties collectively must determine this before the project starts and set up templates to capture the information needed. The structure and level of detail of those templates must be carefully thought out.
5. Train all involved staff in the process of using technology correctly. The staff must be educated on the information you desire, and how to record that information. We suggest that you educate your staff beforehand, provide an example of what a well filled out "report" includes. Also provide a sample of an unacceptable "report" with an explanation of what information is missing and why it is important to include it.

There is a sixth requirement, and it may be the most important, and the most difficult to achieve. The owner and contractor need to change the culture. Placing an emphasis on the kind of detailed relevant reporting described previously is part of changing the culture, but it is only a manifestation of the real and philosophical change needed at construction sites across the country.

The real change is for all parties to be committed to maximizing productivity every day on the job. Every party should be part of the process of scheduling, and of data collection throughout the duration of the process—whether they are specifically required to complete a daily report or not. When any party sees that productivity is suffering, it is every party's opportunity and responsibility to bring the problem to the attention of the other parties. In the past, there have been many construction projects where all the parties involved pointed the finger at each other.

Other than lining the pockets of the attorneys, what was accomplished by such confrontational behavior? In most cases, none of the parties was made whole, particularly when one considers the time devoted to the litigation (rather than devoting that same amount of time more productively to the business of the parties) and the amount of attorney fees and expert fees paid as part of the lengthy legal battle. We fervently believe, (including the author who is an attorney), that it is far better for every party in the construction process to be part of achieving the greatest productivity possible, speaking with candor whenever they detect a diminishment of that productivity, and immediately working together to create a rational solution. A problem solved, or at least addressed, during the construction process is almost always a better and more profitable solution than a legal battle after the project has been finished (which may also sadly describe the state of one or more of the companies involved, their ability to pay and/or their reputations).

There are companies that offer software and services geared to assisting in the improvement of project productivity. During the writing of this book, we contacted one such company and explained what we were doing and asked them for any and all available information. We explained that we would not be buying their software, but we would gladly include any useful information about them in the book. They promised to send information—but never did. We did start receiving marketing emails telling us how great they were. Unfortunately, we are unable to comment on the worth of their product. We do believe that the owner, the agency CM, the contractor, and the subcontractor know far better what information is important to you and to your demonstrated productivity.

4
The Measured Mile

It is encouraging that when claims arise concerning inefficiency, there is an increased use of the objective technique of the measured mile to assess the magnitude of the loss of efficiency. In the past, too often only industry publications or variations of total cost were used to quantify losses associated with productivity. The industry should encourage the use of the measured mile while maintaining the correct application of the method.

This chapter addresses the use of the measured mile as a quantitative tool to determine losses of efficiency on a construction project. It is best to begin with some history concerning the measured mile.

As far as the authors can determine, the first publication related to the measured mile was authored in 1985 and appeared in *Cost Engineering Journal.* Dwight Zink was the author of the article. The article described the measured mile as a method preferred over a total cost approach. It described the measured mile as "comparing the unit productivity costs in an unimpeded time period (or physical area) to that achieved in the claimed disrupted time period (or areas)." The article further described the method as one of comparing man-hours versus percent complete for different periods of time. In its simplest form, it described a comparison of a period of work that had no outside factors affecting progress with a period where acceleration or other detrimental factors were present. This simplistic description is an accurate summary of what a measured mile should be. It should be noted that the use of percent complete or costs might be a misleading element in the use of a measured mile. This will be discussed further. However, for the present, it is most important to understand that a measured mile on a construction project is a quantitative comparison of work with no external negative impacts to work that experienced external factors that could have adversely affected the efficiency of the work.

Ted Trauner, Chris Kay and Brian Furniss. Productivity in Construction Projects, (33–62) © 2022 Scrivener Publishing LLC

The productivity data for both comparison periods is based on the actual project records.

There are many ways that a measured mile may be constructed. Invariably, the ability to utilize a measured mile, and the manner that it is applied are directly dependent on the amount and quality of the documentation available concerning the actual progress of the work.

The earliest uses of a measured mile focused on comparisons of percent complete or costs between the two periods of time being used. There is an inherent problem in using either percent complete or costs. When using percent complete, unless the item being measured is a unit type of item such as feet of pipe installed or yards of fill, then the assignment of percentage completion is a subjective estimate. Seldom do project personnel count exact quantities of an item in assigning a percent complete to a specific line item on the project schedule of values. Consider percent complete on electrical rough-in of a building. At best it may be done by floors or areas. However, the amount of effort may vary among the floors or areas and the percentage assigned will, of necessity, be subjective. When utilizing costs of a line item, this type of measure may also suffer from some degree of inaccuracy since the costs will include materials, subcontractor/supplier invoicing, and other factors that preclude an actual measure of productivity. Also, since some degree of front-end loading often occurs on a construction project, this may dramatically affect the accuracy and results of any measure based on costs or percent complete. As a consequence, using either percent complete or costs as the unit for comparison may lead to erroneous results.

Let's reflect on what our ultimate goal is in the use of a measured mile. We are trying to measure the difference in the demonstrated productivity between a period of time on a project when no external factors or "changes" are affecting the work and a period of time when external factors may be affecting the work. We desire to use a measured mile for several reasons. First, by using actual demonstrated productivity on the project as a baseline we are showing what production was actually achieved with the actual equipment, manpower, and supervision that existed on the project. We are not comparing productivity to some theoretical project or to our estimate that has not been demonstrated on the project. It should be obvious that the significance of the measured mile, if applied correctly, is that we accept as a baseline a productivity rate that has been proven to exist on this job under real conditions. This is a key element of the value of the measured mile and a very important reason to use it if at all possible. The second reason we desire to use a measured mile is that our impacted period or the time when the external factors or changes occurred is also based on a measured production rate with the actual resources applied on the project. Again, we avoid any theoretical measures. Third, the use of a measured mile allows us to show what we actually could do irrespective of estimates, standards, and rates from manuals, etc. But the use of the measured mile demands that we carefully and accurately construct the model in order to accurately portray any impacts that might have occurred to the productivity on the project. Let's explore this a bit further.

One of our first concerns is our unit of measure. The most desirable unit of measure is a productivity unit. A productivity unit would be of the nature of cubic yards per man-hour, cubic yards per crew hour, feet of pipe per crew day, square feet of concrete finished per man-hour, tons of steel set per crew day, feet of pile driven per crew day, etc. In other words, we should look for productivity units that actually demonstrate the work achieved for a given task. Cost per cubic yard is not a productivity unit—it's a unit cost measurement. Percent complete per day is not a true rate of the work achieved for reasons previously stated. In order to construct a measured mile with a true productivity rate, we must maintain records that allow for productivity to be determined. Recognize that not all tasks are finitely measurable such that a pure productivity rate can be documented. Accepting that, we want to determine as many tasks that can have their productivity measured as possible and use as many of these as feasible.

A second concern in the use of the measured mile is the determination of the periods that are to be measured. In the purest form, the comparison for a measured mile should be between a period of work that has no adverse factors that may negatively affect the work. In this context, adverse factors are those elements that are being claimed as a cause of problems or a reduction in efficiency. Clearly, poor supervision may be considered an adverse factor but if this factor is present in both periods of the comparison, it equalizes out. For example, if a contractor alleges that excessive overtime that was caused by the owner led to reduced efficiency, the baseline period for the measured mile should be a period when either no overtime was worked or the original planned overtime was worked. The second period for the comparison would likely be a period when the contractor was working excessive overtime. It is possible that there may be multiple periods in the comparison. Staying with our example of excessive overtime, let's assume that a contractor was working 40-hour weeks, five days at eight hours per day. The owner directs overtime, and the contractor goes to 50 hours per week, or five days at 10 hours per day. The owner later directs even more overtime, and the contractor goes to 60-hour weeks, six days at 10 hours per day. In this example, the baseline period would be the 40-hour week. A comparison would separately be made with the 50-hour week period and the 60-hour week period. The two comparisons may yield different measures of reduction in efficiency.

With respect to the periods chosen, the baseline is best established as a period when no adverse factors are occurring. But what if no such period exists? Instead, what do we do if some factors occur throughout the project while other factors occur sporadically? In these cases, the baseline measure may be diluted to a period of least impacted or least affected productivity. In other words, the baseline period is the period of work when it was as good as it could be but was still affected to some degree. This measure would conservatively favor the owner since the measure of inefficiency could be higher if no factors are present.

Another very important element of the measured mile approach is the productivity unit that is being used for the comparison, in other words, the type of work that is chosen. The analyst should strive to compare "apples to apples." For example, if the measure is based on concrete placement, the type of placement should be the same. Comparing concrete columns to concrete slabs would not be a valid comparison since one would reasonably anticipate that the cubic yards per man-hour for slab placement would be significantly higher than the cubic yards per man-hour for columns. Some analyses attempt to take a generic item such as concrete and presume that placement of concrete is a valid measure with no thought or reference to the type of members that are being placed. This is erroneous and will yield suspect results. The same line of reasoning applies to other types of work. For example, feet of pipe per crew day. If one pipe is 72″ diameter and the other is 16″ diameter, there is not a valid correlation between the productivity between the two. At times you may see an analyst attempting to "normalize" different pipe sizes to allow the 72″ to 16″ comparison. In most cases it is questionable if this normalization can be done accurately. The authors reviewed a "measured mile" presented during the course of litigation on a public works project that involved laying pipe in a city. The analyst used a measure of feet of pipe per day. No distinction was made for size of pipe, depth of pipe, or size of crew. When the presentation was scrutinized, it became clear that these oversights were significant to the results. On some days, the crew laying pipe was four laborers and on other days the crew was 20 laborers. Obviously, just the crew size difference affected the amount of pipe that was laid. Likewise, some days only 24″ pipe was placed and other days only 72″ pipe was placed. It was no surprise that the productivity for the 72″ pipe was significantly less than for the 24″ pipe. In this example, the input unit chosen, feet of pipe per day, represented production and was too general and too broadly defined to provide a reliable measure or comparison of productivity. Both the unit chosen and the type of work were inconsistent with the proper application of the measured mile.

To provide another example, the authors also reviewed a measured mile analysis of welding productivity for structural steel members. The alleged measured mile performed attempted to quantify welding productivity rate differences when welding rejections increased on the project. Again, the analysis ignored key factors affecting the expected productivity between the two measured time periods, including variances to material type, welding process, expected rejection rates of the different welding process, and other factors. Ignoring other flawed implementation steps, the results compared two periods when the work was "apples and oranges" instead of "apples to apples." No distinction was made for the expected productivity differences during the time period measured and, as a result, there was not a valid measured mile comparison for productivity losses caused by welding rejections.

Calculating Productivity Losses and Inefficiency Using the Measured Mile

Let's go back to our example of drainage pipe placement to demonstrate how to perform the measured mile calculations. We will change the facts slightly to simplify the calculation, and then build on the fundamentals to demonstrate variants of the calculations.

First, the contractor performed installation of 36" reinforced concrete pipe (RCP) during a time period unaffected by owner changes (unimpacted period). Presume the pipe size, depth, bedding, processes, and other variants were similar between the unimpacted and impacted periods in order to keep things simple. The contractor's daily reports demonstrated that the contractor placed 880 linear feet (LF) during a two-week period without any owner impacts. The crew that placed the pipe consisted of 3 laborers, a backhoe, and a backhoe operator working nine hours per day. The labor hours reported on the contractor's daily logs for this work was 360 labor hours (4 workers × 9 hours a day × 5 days a week × 2 weeks = 360 hours). The unimpacted productivity was:

$$Unimpacted\ Productivity = \frac{Outputs}{Inputs} = \frac{880\ LF}{360\ Hours} = \frac{\mathbf{2.4\bar{4}LF}}{\mathbf{Hour}}$$

The crew operated at that average level of productivity for two weeks during the unimpacted period before an owner-directed change occurred that caused the crew to work 60 hours per week (5 workdays × 12 hours = 60 hours) for the next two weeks (impacted period). During the impacted period with increased overtime, the contractor's documents showed that it placed 960 LF, which was an increase compared to the prior period, and expended 480 labor hours completing the 36″ RCP. The impacted productivity was:

$$Impacted\ Productivity = \frac{Outputs}{Inputs} = \frac{960\ LF}{480\ Hours} = \frac{\mathbf{2.00\ LF}}{\mathbf{Hour}}$$

Comparing the productivity rates of the impacted and unimpacted period yields the following productivity loss:

$$Productivity\ Loss = Unimpacted\ Rate - Impacted\ Rate$$

$$= \frac{2.4\bar{4}\ LF - 2.00\ LF}{Hour} = \frac{\mathbf{0.4\bar{4}\ LF}}{\mathbf{Hour}}$$

The measured mile allowed the contractor to understand that during the impacted period, its productivity decreased by 0.44 LF per hour. To calculate the inefficiency, the contractor compared the productivity lost during the impacted period to the productivity expected during the unimpacted period:

$$Inefficiency = \frac{Productivity\ Loss}{Unimpacted\ Rate} = \frac{0.4\overline{4}}{2.44} = \mathbf{18.\overline{18}\%}\ \boldsymbol{Inefficiency}$$

In summary, when the contractor worked 60 hours per week during the impacted period, it was 18.18% less efficient than when it worked 45 hours per week during the unimpacted period.

The inefficiency occurred during the unimpacted period that had 480 labor hours. Therefore, the inefficiency rate of 18.18% resulted in the following additional hours:

$$Additional\ Hours = Total\ Hours\ during\ Inefficient\ Period\ x\ Inefficiency\ Rate$$

$$= 480\ hours\ x\ 0.18\overline{18} = \mathbf{87.\overline{27}}\ \boldsymbol{additional\ hours}$$

An easy verification of the 87.27 additional hours would be to take the inverse of the productivity ratio and express the ratio in terms of hours per linear foot of pipe. This would result in the following unimpacted and impacted productivities:

$$Unimpacted\ Productivity = \frac{Inputs}{Outputs} = \frac{360\ Hours}{880\ LF} = \frac{\mathbf{0.\overline{409}}\ \boldsymbol{Hours}}{\boldsymbol{LF}}$$

$$Impacted\ Productivity = \frac{Outputs}{Inputs} = \frac{480\ Hours}{960\ LF} = \frac{\mathbf{0.5}\ \boldsymbol{Hours}}{\boldsymbol{LF}}$$

$$Added\ Hours\ Per\ LF\ during\ Impacted\ Period = 0.5 - 0.\overline{409} = \frac{0.\overline{09}\ Hours}{LF}$$

$$Added\ Hours\ for\ Impacted\ Period = \frac{0.\overline{09}\ Hours}{LF}\ x$$

$$960\ LF\ for\ Impacted\ Period = \mathbf{87.\overline{27}}\ \boldsymbol{Added\ Hours}$$

The preceding calculation provides a verification of the first calculation method and also demonstrates an alternate way to calculate the additional hours expended during the impacted period.

To calculate the inefficiency costs, presume the average cost of all straight-time and overtime hours during the unimpacted period was $40 per hour. The additional cost is shown in the following equation:

$$\textit{Added Labor Cost during Impacted Period} = \$40 \textit{ per hour x } 87.\overline{27} \textit{ added hours} = \mathbf{\$3,490.91}$$

The determination of burden and other markups would be added to the preceding amount and may require other information on the contractor's cost-accounting records. For example, many companies attribute burden to only straight-time hours and not to overtime or double-time hours. As a result, the additional burdened cost requires further adjustment of the additional hours to account for only the additional straight-time hours before calculating the additional burden cost.

Calculating equipment inefficiency costs may also require more detail than labor calculations, especially when the equipment costs occur during non-straight-time periods, or the equipment is owned instead of rented. For example, equipment may not incur an hourly rental cost beyond eight hours per day, except for the operating cost (equipment, fuel, lube, operator, etc.). Owned equipment costs may be further complicated by the contractor's internal accounting procedures and require detailed equipment cost accounting to quantify the equipment costs due to inefficiencies. This again highlights the importance of separating the inefficiency calculations from cost calculations, as the two calculations require the consideration of different variables in order to obtain the proper results.

Note that the inefficiency calculation was supported by the project records on the actual project where the inefficiency occurred. This allowed for an inefficiency analysis using the measured mile method. However, what if the project is adversely affected throughout the entire period of the job? In other words, there is no time period that reasonably established the efficiency that could have been achieved except for the adverse factors. In these cases, the next best measure that can be used would be a comparison to other similar projects.

Comparisons to a Similar Project

What are "similar projects?" A similar project would be one that is the same type of construction such as paving a roadway. The comparison to a similar project is the next-best method to measure inefficiency if a measured mile cannot be performed. The decrease in preferability results from the increase in variabilities between the source project with the inefficiency and the "similar" project being used as the basis of the unimpacted period. To reduce variation among the comparison projects, it should be in the same geographical region so that weather, labor factors, and regional influences are similar. It should be by the same contractor, or subcontractors, performing the work. It should have the same or similar management staff. The execution of means and methods, equipment used, etc., should be similar. The general makeup of the job or at least the main elements being compared should be similar. For example, paving an econocrete airfield runway would not be "similar" to slipform paving a concrete roadway. Placing Portland cement concrete pavement in Arizona may have different productivity considerations than placing similar material in New Jersey. While comparisons to similar projects are less preferred than the measured mile, there are occasions when a project does not have a period of time that can be identified as unimpacted or least impacted period of productivity. When this approach is used, a careful review must be made to make sure the projects are similar and to carefully review the projects for any differences that may factor into the results of the analysis.

The change from a measured mile to a comparison project does not require changes to the inefficiency calculations, as the inefficiency calculations are the same regardless of whether the measured mile or similar project methods are used. What changes is that the basis for the unimpacted period shifts from the source project to the comparison project. As discussed earlier, this apparent subtle change to the source of the unimpacted period data means that the analysis needs to consider and potentially reconcile more variables that affect productivity between the source and comparison project.

Measured Mile Case Studies

To best describe the approach to the measured mile, several case studies have been included in this chapter. The case studies are from real projects and cover a wide variety of types of work. At the end of each case study a commentary is provided that addresses the strong and weak points of each presentation. The authors believe that the case study method is the best approach to exemplify the application of the measured mile technique and some of the considerations that should be made during a review of a measured mile analysis.

The one most significant point that the reader should take away from this chapter and the case studies is that a measured mile can best be done when good records exist. This means that contractors and owners alike should structure their recordkeeping such that it provides measurable productivity units or the ability to derive them. Project documentation must record man-hours spent, and quantifiable units of work accomplished by specific type, size, and quantity of material. Realistically, recordkeeping of this nature is not as difficult as some may think. If mechanisms are set up at the beginning of the project, accurate tracking of the necessary information becomes a routine task that can yield benefits not only in the case of a dispute concerning loss of efficiency, but also in terms of providing useful information during the project and for future projects and bids.

Case Study #1: Transportation Project

In the course of this project, the contractor worked overtime throughout the project in order to reduce the delays allegedly caused by the Department of Transportation. It is generally acknowledged that prolonged or extended periods of overtime work may result in a reduction of productivity. The available project data was reviewed to determine if a measurable loss of efficiency occurred.

The preferred approach to the analysis of inefficiency is called the "measured mile." The measured mile approach relies on a comparison of productivity before (the unimpacted period) and after (the impacted period) an impact occurs. Such an analysis would be appropriate to address the effect of the extended periods of overtime on the contractor's productivity.

In order to evaluate the impact on the contractor's productivity due to overtime using the measured mile approach, it was necessary to identify an activity that included a repetitive task performed both before and after the extended overtime and for which productivity data was available. The activity that met these criteria was the concrete paving work.

According to the plans, an 11-inch plain concrete base pavement was to be constructed for the proposed roadway throughout the project. The as-built information indicated that the majority of the concrete base pavement was placed with a slipform-paving machine. In order to perform the measured mile analysis, the contractor's paving productivity was compared before and after the extended overtime period.

The first step in this analysis was to identify the production data for concrete paving activities. According to the available project documents, Bid item 11″ Concrete Base, corresponded to this work. The units of work for this item were measured in square yards of concrete pavement. Next, it was necessary to establish a basis of measurement for the paving operations. Because the crew sizes for the concrete pavement construction were approximately the same and because the daily logs listed labor hours for the entire crew, the duration of the concrete base pavement work was measured in crew hours. Therefore, productivity was defined in terms of square yards of concrete pavement per crew hour (SY/crew hour).

To facilitate this analysis, the repetitive slipform construction of the concrete base pavement was evaluated. For this analysis, the actual crew hours spent on the slip-formed pavement construction were tabulated. Crew hours spent on other concrete roadway construction activities such as "hand paving" or the placement of concrete traffic barriers were not included in this analysis.

Next, an unimpacted period for the concrete paving activity was identified and is referred to as the unimpacted period. According to the daily logs, the concrete paving crews during this time period consisted of superintendents and skilled and unskilled laborers. Neglecting the superintendents or shop stewards, the manpower averaged 22 men per crew. According to the as-built information, during the unimpacted period, the contractor slip-formed 6489.2 S.Y. of eleven-inch plain concrete base pavement, expending 45 crew hours. Therefore, during this unimpacted time frame, the contractor demonstrated a productivity rate of 144.2 S.Y. per crew hour (6489.2/45=144.2).

Next, the productivity for the concrete paving work was evaluated for a time period impacted by prolonged overtime operations. During the impacted period, the contractor went from working a five-day work week with nine-hour days to a six-day work week with ten-hour days. Although the concrete crews consistently worked extended hours during the impacted period, the contractor worked exclusively on the 11-inch base pavement operations using a slipform paver.

During the impacted time period, the manpower also averaged 22 men per crew for concrete paving. According to the as-built information, the contractor paved 6893.9 S.Y. of 11-inch plan concrete base pavement using a duration of 58.5 crew hours. Therefore, during the impacted time period, the contractor realized a productivity rate of 117.8 S.Y. per crew hour (6893.9/58.5=117.8). Thus, after working four consecutive weeks of extended overtime operations it appeared that the contractor's productivity for slip-forming 11-inch concrete base pavement decreased from 144.2 S.Y. per crew hour to 117.8 S.Y. per crew hour.

In order to calculate the inefficiency caused by the extended overtime work the productivity for the impacted period was divided by the productivity of the unimpacted period.

$$\frac{117.8\ S.Y.\ of\ concrete\ base\ pavement\ per\ crew\ hour}{144.2\ S.Y.\ concrete\ base\ pavement\ per\ crew\ hour}\times 100\% = 81.7\%$$

Therefore, the inefficiency caused by the extended period of overtime was 18.3% (100%-81.7%=18.3%).

What lesson was learned from this situation? A party cannot articulate the "measured mile" approach unless it has data for the unimpacted period and the impacted period. Here, the data compiled on the project included the number of square yards of concrete base pavement, the number of crews, and the number of hours expended per employee. As a result, the productivity could be calculated for both the unimpacted and impacted periods of time, making it possible to perform a "measured mile" approach to quantify losses in productivity.

Case Study #2: The Pharmaceutical Facility

The project involved the construction of a pharmaceutical facility in the approximate amount of $100,000,000. During the project, a significant number of changes were made by the owner through formal change orders, drawing revisions, sketches, etc. Overall, the project had 2,000 changes at a minimum. Needless to say, the project ended in a dispute that was ultimately resolved during mediation. The specific problems identified included extra work, delays, and loss of productivity. This case study highlights one small portion of the project, focusing on the lost productivity for the mechanical contractor.

In order to measure the lost productivity, an analysis was made on the ductwork installed by the mechanical contractor. The project had nine buildings broken down into three areas. Area 1 included Buildings 1, 2, and 3. Area 2 included Buildings 4 through 7, and Area 3 included Buildings 8 and 9. Area 2 had the best available information for analytical purposes. In that area, Building 5 had the fewest changes and disruptions. This does not mean that Building 5 did not have problems but that it had the least amount of problems. As a consequence, Building 5 was used as the least impacted building for the analysis.

Based on the project records, Building 5 had 201,736 pounds of duct installed. The labor hours for this installation totaled 14,228 man-hours (MH). This resulted in a demonstrated productivity of 14.18#/MH (201,736# divided by 14,228 MH). Based on similar calculations the other buildings in area 2 had the following levels of productivity:

Building 4 - 10.06#/MH
Building 6 - 10.61#/MH
Building 7 - 11.82#/MH

Based on the demonstrated productivity in the least impacted building, the losses for the other buildings in Area 2 were calculated as follows:

Building 6: Total pounds of duct = 312,577#

$$\frac{312{,}577\#}{14.18/MH} = 22{,}044\ MH$$

Actual man-hours expended = 29,461 MH
Extra man-hours:

$$29{,}461\ MH - 22{,}044\ MH = 7{,}417\ MH$$

Similar calculations were performed for Buildings 4 and 7 and resulted in a total loss of man-hours of 17,735 MH. The total man-hours expended for these three buildings was 70,268 MH. Therefore, the percentage loss of efficiency is calculated as:

$$\frac{17{,}735\ MH}{70{,}268\ MH} = 25.24\%$$

A second analysis was performed adjusting for any approved change orders. Basically, the approach was the same, but the pounds of duct and man-hours were decreased for any changes that could be documented. This resulted in a loss of efficiency of 23.83%. To be conservative, the smaller percentage was used for the calculation of damages.

Based on the analysis for Area 2, the 23.83% loss of efficiency was also applied to Area 1, Buildings 2 and 3.

The ductwork in Area 3, however, was performed by another contractor. While the problems were the same and at least as significant as Area 2, the contractor for Area 3 calculated its loss of efficiency based on its estimated ductwork productivity versus its actual productivity. This approach yielded a loss of approximately 20%. To be conservative, the lesser figure of 20% was used for the loss of efficiency in Area 3.

Other approaches were also used in order to verify the reliability of the loss calculated for the ductwork. An analysis of the piping resulted in a calculated loss of 21.19% and this figure was used for the applicable man-hours. Similarly, several other analyses were run on the different areas and buildings. The end result was that all approaches resulted in very close agreement with the initial calculations.

In the presentation of the costs associated with the loss of productivity, all calculations were shown so that all parties could make a reasoned decision as to the most acceptable method and number to resolve the dispute.

In this instance, the availability of good records was essential to the analysis. It must be noted, however, that a significant amount of work was required to determine pounds of duct, feet of pipe, etc., in order to perform the analysis. Hence, we must be as detailed as possible when we structure the recording of our productivity.

Case Study #3: The Psychiatric Hospital

This project involved the construction of a County Psychiatric Hospital. This included the construction of a new pavilion and renovation of two existing structures. The electrical contract was in the approximate amount of $5,000,000.

During the course of the project there were numerous design changes made, multiple errors and omissions in the plans, problems with the work of other subcontractors, and a constant array of clarifications and modifications. As a result of this, the project was delayed and was not performed in the orderly sequence originally planned. As a result, a dispute arose and required an analysis of losses in productivity along with several other issues. The following presents an overview of the analysis utilized to measure the loss of productivity by the electrical contractor during the construction.

Based on a review of the project documents, it appeared that the majority of the losses in productivity occurred in the branch conduit installation. Consequently, that area became the initial focus of the analysis.

The project information indicated that the first few months of branch conduit installation were performed with minimal problems. To verify this, branch conduit work was plotted on a daily basis to determine if the work proceeded logically and methodically in accordance with the project schedule and plan. The as-built information/plot verified that this work did proceed in a reasonable fashion for the first few months. Thereafter, the work was performed in a disrupted and sporadic manner corresponding to the problems that were occurring on the project. However, the first six months allowed a representative period to establish a demonstrated level of productivity of the electrical contractor's crews.

The project documents supported that the electrical contractor was able to install 45% of the branch conduit during the first six months on the project. This totaled 107,525.25 feet of conduit with a total of 6,957 man-hours (MH) expended for the installation. This calculates to a productivity rate of 15.45 feet/MH.

For the remainder of the project, that portion of the project impacted by the changes, the electrical contractor installed 59,736.25 feet of branch conduit with a total of 15,048.75 man-hours expended for the installation. This calculates to a productivity rate of 3.97 feet/MH.

Based on a comparison of the unimpacted period with the impacted period of work, the electrical contractor expended 10,823 MH in excess of what it should have because of the lost productivity. This is calculated as follows:

Demonstrated rate of productivity = 15.45feet/MH

Actual quantity installed during impacted period = 59,736.25 feet
Man-hours expected: 59,736.25 feet/15.45feet/MH = 3,866.42 MH
Actual MH expended = 15,048.75
Overrun due to loss of productivity = 11,182.33 MH

Similar calculations were performed for fixture installation work.

As can be seen from this example, the ability to determine the effort expended in specific tasks is key to being able to determine productivity rates and draw conclusions or "measure" any differences in productivity on a construction project.

Clearly it would have been desirable to have even more detailed information in order to refine the analysis. Having such contemporaneous recordkeeping may well have allowed for fewer overall problems on the project.

Case Study #4: Pile Driving

In this project, a state Department of Transportation (DOT) contracted for the construction of a new highway bridge. The bridge had 25 separate piers and each pier was constructed on pre-cast pre-tensioned concrete piles. During the pile driving operation, the contractor asserted that a differing site condition had been encountered that adversely affected the pile driving operation. The contractor specifically stated that boulders and rubble had been encountered that obstructed the pile driving. Upon review of the soil borings for the project, several borings indicated the presence of boulders and rubble. However, these borings had not been included with the contract documents and also had not been made available to bidders. Consequently, it appeared that the contractor had encountered a differing site condition. The claim that was submitted by the contractor included delays to the project and lost productivity, both directly related to the pile driving operations.

The DOT had very detailed records of all of the pile driving operations. The particular inspector for this phase of the project had tracked the driving time for every pile to the minute. He also had recorded the time for splices, moving equipment, the efficiency of the hammers, the length of every pile, the time for cutoffs, length of cutoffs, etc. In fact, the records were the best our staff had ever seen on a pile driving operation. The detailed records allowed for a precise analysis of the time for each pier and each pile.

Based on these detailed records, the DOT was able to accurately respond to the claim. The contractor had asserted that the operations were affected on eight piers because of the differing site condition. He also asserted that in other piers, it had substantial cutoffs and also had piles go much deeper than planned. The records allowed the analysis to categorize the piers into four separate groups. The first group was piers that did not experience a differing site condition and the production pile lengths did not vary significantly from the test pile length. This group included 12 piers and the piles in these 12 piers averaged a driving rate of 3.4 piles per crew day. The second group of piers included piers that experienced a differing site condition but did not have production pile lengths that varied significantly from the test piles. This group included five piers and the demonstrated productivity was 3.8 piles per crew day. The third group of piers was those that experienced a differing site condition and the production pile lengths varied significantly from the test pile lengths. This group had six piers and the demonstrated productivity was 2.9 piles per crew day. The final group included piers that did not experience a differing site condition, but production pile lengths varied significantly from the test pile lengths. This group had two piers and a demonstrated productivity of 4.4 piles per crew day.

Based on the data and the actual pile driving records, it was relatively simple to determine those piers that were affected by differing site conditions and quantify the effect to the contractor. From that, reasonable compensation could be determined.

This project is an outstanding example of why detailed records are important and the level of detail that should be recorded for operations such as this.

Case Study #5: DOT Project

A contractor was executing the work on a contract with a state Department of Transportation (DOT). During the course of the work numerous changes were made that adversely affected the project schedule and efficiency. The contractor submitted a claim for the additional costs. Summarized below is the contractor's presentation of a measured mile for the embankment work on the project.

As stated by the contractor in the claim, there were few work items or time periods that were not impacted in one way or another. However, the

DOT project productivity comparison on embankment work.

Year	Cost code	Description	Quantity (CY)	Man-Hours (MH)	Productivity (CY/MH)
1	150	Embankment (APP)	133,501	4,258.5	31.35
1	135	Embankment	12,000	362.5	33.10
2	135	Embankment	169,511	13,412.0	12.64
3	135	Embankment	54,640	7,023.0	7.78
4	135	Embankment	60,411	3,788.5	15.95

contractor was able to identify two work items where there existed a period of time where relatively few impacts occurred and that the impacts of overtime could be addressed. One of these items was the embankment work.

The contractor's analysis selected the first year of embankment work and compared it to the succeeding three years of embankment work. This first year of embankment required a minimum amount of overtime while the second two years of work required significantly more amounts of overtime over a much longer period. The contractor presented the following data from his cost accounting system:

Based on the preceding data, the contractor ignored cost code 135 for the first year since the cost code of 150 was a more conservative number. Hence, it used the demonstrated productivity of 31.35 cy/mh as the demonstrated productivity.

On a percentage basis, the calculations show the following:

Year	Inefficiency calculation		Inefficiency factor
2	(12.64 – 31.35)/31.35	=	59.7%
3	(7.78-31.35)/31.35	=	75.2%
4	(15.95-31.35)/31.35	=	49.1%

To be conservative, the contractor then calculated inefficiency factors for the bridge work for three railroad bridges on the project. This analysis was more complex because of the number of work items for the three railroad bridges. Consequently, these detailed calculations will not be presented herein. The bridge analysis yielded an inefficiency factor of 22.6% and that factor was used for the subsequent cost calculations.

In general, the approach used by the contractor is reasonable. Some study is warranted for the exact differences in the cost codes used and the weighted average based on man-hours that the contractor used for the composite bridge work. Similarly, the cost accounting should be scrutinized to ensure that the data is reasonably accurate and reliable for a comparison of like items. Once again, the more detailed the records, the more reliable the calculations become.

Case Study #6: Welding Work

Project Background

A structural steel subcontractor was awarded a subcontract to fabricate, weld, and ship approximately 20 different structures for light rail stations. The subcontract agreement was based on a time and materials (T&M) basis, meaning that the subcontractor was reimbursed for the time and materials expended to complete the work. Each of the structures required the fabrication of the steel components, and were followed by the assembly, welding, and quality control and assurance work, and then the shipping of the product to the project site. The steel was accepted on-site and erected by a separate subcontractor at the site that was unaffiliated with the steel supplier. As a result, all of the work to be completed by the steel subcontractor was completed in its fabrication facility, and all of the quality control and assurance work was also completed in the subcontractor's facility. The general contractor employed a separate steel inspector to perform reviews of the steel subcontractor's work, and the inspector provided daily and monthly reports to the general contractor that, in turn, were provided to the project owner. The inspector's reports were not provided to the steel subcontractor.

Welding Types and Methods Used

The subcontractor started his work by completing a mock-up of one of the structures, which was later approved. The hours expended by the subcontractor to complete the mock-up, along with the linear footage of welding completed, was tracked by the subcontractor.

Following the completion of the mock-up, which was later used as one of the structural components, the steel subcontractor proceeded with fabrication and welding work on the remaining structures. The type of weld used to weld the structures (complete joint penetration (CJP), partial joint penetration (PJP), and fillet welds) depended on the structural requirements of the project. All structures required a mix of CJP, PJP, and fillet weld types. However, the welding method used to complete the weld type was a means and method selection by the steel subcontractor. The steel subcontractor manually welded, or hand welded, the first several structures using a mixture of flux cored arc welding (FCAW) and submerged arc welding (SAW) methods to complete the various welds.

Approximately mid-way through the remaining structures, the steel subcontractor changed to predominantly using machine welding via gas metal arc welding (GMAW) method to expedite the welding process, and then used hand welding processes only when necessary. In summary, the first half of the work was completed using predominantly a hand-welding method, while the second half was completed using predominantly machine welding. The first half of the structures consumed a higher quantity of hours than the second half of the structures. Once the machine welding started for the second half of the structures, the rate of linear feet welded per man-hour increased. This resulted in an improved productivity rate per linear foot of welding during the machine welding.

Each weld was inspected via ultrasonic testing (UT) to ensure the structural integrity of welding. The inspections identified welding issues requiring corrections during both the hand and machine processes. However, despite the increased productivity of the machine welding process, there was also an increase in welding rejections using machine welding. Each weld rejection required remedial work to repair the area with the apparent welding issue, along with additional re-inspections of the weld using UT. It should be no surprise that the remediation work and re-inspection process consumed additional time.

The Dispute and Claim

When the steel subcontractor was approximately 95% complete, the general contractor (GC) notified the steel subcontractor that it was withholding the remaining subcontract balance and all retainage amounts. The GC stated that the steel subcontractor's chosen means and methods resulted in additional time expended to remediate the non-conforming welds, and that the general contractor should not have to pay for remedial work. The GC also backcharged the steel subcontractor for some of the amounts paid to the inspector, as the GC alleged that the inspector's costs were increased due to the increased non-conformances during the second half of the work.

To support its claim against the steel subcontractor, the GC's expert completed an inefficiency analysis. The GC's expert used various measurements of productivity during the project to complete its "measured mile" analysis, and based on the expert's analysis, the second half of the project incurred productivity losses due to the increased rate of non-conforming welds and the subsequent remedial work. A second expert for the GC reviewed and verified the work of the first expert, and both testified that the subcontractor's costs increased during the 2nd half of the project due to the non-conformances, which caused unnecessary cost increases for the GC.

Other Experts' "Measured Mile" Analysis

As already stated, the subcontractor tracked the hours used to complete the work. The welding inspection reports allowed for an understanding of when the various welds were inspected and, if necessary, re-inspected. The frequency of welding inspections varied between a daily and weekly inspection ratio. The experts assumed that the linear footage of welds inspected equated to the linear footage of welding completed since the last inspection. For example, if a welding inspection occurred on Days 7 and 11, it was presumed that the Day 7 inspection included the linear footage of welding completed on Days 1 through 6, while the linear footage of welding completed on Days 7 through 10 were included in the Day 11 inspection. This allowed the experts to match the linear footage of welding completed (quantity) with the hours expended to produce that quantity as long as all the hours from Days 1 through 10 were used to calculate the productivity (more on this later).

The experts' "measured mile" analysis used a Baseline Period that spanned between mid-August and mid-September 2018. This Baseline Period provided the basis for the "unimpacted" time period or, in other words, the productivity rate that the steel subcontractor should have been able to weld throughout the rest of the project. The hours of welding per month are shown in Figure 4.1, along with the Baseline Period.

In total, the experts' Baseline Period spanned 25 working days. Of the total of 25 working days during the Baseline Period, inspections occurred on 11 working days. The experts showed that the welding rejections during this time period were the lowest on the project compared to the other months. The experts used the labor hours expended on those 11 working days and the linear footage of welding inspected on the 11 working days to deduce the productivity rate for welding completed. The Baseline Period productivity is shown in the following equation:

$$\frac{301\,Linear\ Feet\ of\ Welds\,Completed}{673\,Labor\ Hours} = \frac{0.447\,LF}{1\,Hr.}\ Baseline\ Productivity\ Rate\,(Unimpacted)$$

The experts asserted that since the welding non-conformance rate was the lowest during the Baseline Period, the rest of the project should have had the same productivity rate as the Baseline Period.

The experts then compared the productivity during the Baseline Period (Unimpacted Period) to the productivity during the remainder of the

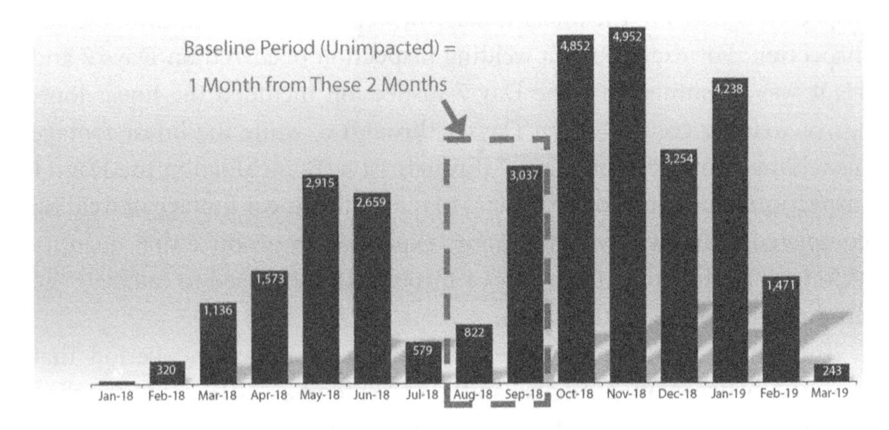

Figure 4.1 Hours of welding per month along with Baseline Period.

project (Impacted Period). The productivity that occurred during the Impacted Period is shown in the following equation:

$$\frac{10{,}573 \text{ Linear Feet of Welds Completed}}{30{,}646 \text{ Labor Hours}} = \frac{0.345 \text{ LF}}{1 \text{ Hr.}} \text{ Impacted Productivity Rate}$$

The productivity loss is calculated showing two different methods:

Method 1 – Using Productivity Rates as Shown by Linear Feet/Hour

$$\frac{Unimpacted - Impacted}{Unimpacted} = \frac{0.447 - 0.345}{0.447} = 22.8\% \text{ Productivity Loss}$$

Applying productivity loss against the total hours during the Impacted Period would result in the following additional hours:

$$22.8\% \text{ Productivity Loss } x \text{ } 30{,}646 \text{ Hours} = 7{,}006 \text{ Lost Hours (Added Hours)}$$

Method 2 – Using Productivity Rates in Hours/Linear Foot

If the productivity was calculated as labor hours per linear foot, the calculations show the same outcome, as shown in the following:

$$\text{Unimpacted Rate} = \frac{673 \text{ Hours}}{301 \text{ Linear Feet}} = \frac{2.236 \text{ Hours}}{1 \text{ Linear Foot}}$$

$$\text{Impacted Rate} = \frac{30{,}646 \text{ Hours}}{10{,}573 \text{ Linear Feet}} = \frac{2.899 \text{ Hours}}{1 \text{ Linear Foot}}$$

$$\frac{Impacted - Unimpacted}{Impacted} = \frac{2.899 - 2.236}{2.899} = 22.8\% \text{ Productivity Loss}$$

Applying this against the actual hours during the Impacted Period yields the same 7,006 added hours as Method 1.

Alternatively, the analyst could use the following equation against the linear feet completed during the Impacted Period:

$$\frac{2.899\,Hours}{1\,Linear\,Foot}-\frac{2.236\,Hours}{1\,Linear\,Foot}=\frac{0.663\,Added\,Hours}{1\,Linear\,Foot}x10{,}573\,Linear\,Feet=7{,}006\,Added\,Hours$$

Since the welding non-conformance rate was higher throughout the rest of the project, the experts alleged that the steel subcontractor should have completed the work in 7,006 fewer hours, resulting in an alleged overbilling of 7,006 hours to the GC. The additional inspector hours were added, multiplied by the respective labor rates, and then marked-up to quantify the alleged damage to the GC.

A Big Problem

You may have caught it by now, but the experts' productivity calculations during the Baseline Period (Unimpacted Period) contained a significant error. Remember that the quantities during 11 inspection days included all welding completed since the prior inspection date. In other words, while the inspections only occurred on 11 of the 25 working days during the Baseline Period, the quantities were actually achieved over all of the 25 working days. This would mean that the labor hours on all 25 working days would need to be used—not just the labor hours on the 11 inspection days. The experts should have used 1,452 labor hours instead of 673 labor hours. Adjusting for this error leads to—you guessed it—a lower productivity rate during the Baseline Period. The original Baseline Period and corrected Baseline Period calculations are shown in the following table, along with the productivity during the alleged Impacted Period:

Original and corrected Baseline Period with productivity.

Time period	Weld length (LF)	Labor hours	Productivity (LF/Hr.)
Original Baseline Period	301	**673**	0.447
Corrected Baseline Period	301	**1,452**	**0.207**
Impacted Period	10,573	30,645	**0.345**

This means that the productivity during the Impacted Period (0.345 LF/Hr) was better than the Unimpacted Period (0.207 LF/Hr) when all labor hours from the Unimpacted Period were accounted for. There was no loss of productivity during the Impacted Period and, appropriately, no damages were awarded by the trier of fact.

Other Productivity Considerations

Other considerations for the inefficiency analysis on this project were:

- The measured mile must compare similar work
  - The fabrication hours were ignored by the analysts and were blended with the welding hours. The problem with this was that fabrication hours were not uniform throughout the welding periods and, therefore, did not substantiate a like-for-like comparison necessary for the measured mile.
  - Weld types were also not uniformly distributed because the material being welded was not uniformly distributed. CJP welds were more complex welds that caused a higher non-conformance rate during the UT inspections. A much higher proportion of CJP welds were completed during the Impacted Period than the Baseline Period. Despite this, the Impacted Productivity was still better, but it's questionable whether this was also a valid like-for-like work comparison.
- Test the strength of the assumed relationship
  - The analysts presumed that an increase in welding non-conformances had to cause an increase in the hours needed to complete each structure. The analysts relied on a practical argument but did not test the strength of that argument using the project data. The analysts could have tested the strength of the assumed relationship by using a regression analysis to compare the number of non-conformances per structure (independent variable) to the hours it took to complete each structure (the dependent variable). The analysts assumed 100% of the added hours were caused by increased non-conformances, but the data showed otherwise. The resulting r-squared value of a simple linear regression was 0.458, meaning that, at

> most, only 45.8% of the variation in welding hours per structure *may* be explained by the changes to the welding non-conformances. In other words, at best, less than half of the additional hours *may have been* caused by the increases in welding non-conformances.

There were more issues, but we'll skip to the answer. The project data showed that the steel subcontractor changed from hand to machine welding to increase the efficiency of the welding process. This change decreased the hours spent per linear foot of weld by almost 50%. Yes, additional hours were expended to resolve the increased welding non-conformances and those non-conformances consumed a small portion of the 50% increase. But, overall, had the steel subcontractor continued the process using the hand welding methods, it would have resulted in increased labor hours billed to the GC under the T&M contract. The steel subcontractor's change from hand to machine welding increased welding non-conformances, but overall, it saved the GC considerable money in increased process efficiencies.

Case Study #7: The Painter

A painting subcontractor was awarded a bid for interior painting of a new high-rise building. The initial (lower) floors of painting progressed with few if any problems. However, as the painter worked himself up the buildings, he experienced numerous interferences in the progress of his work. Other trades were working on floors performing remedial work and, in some cases, still performing base contract work. Similarly, since some of the base contract work was not yet completed, the entire floor was not available for the painter to efficiently accomplish his work. Accordingly, the painter's work on at least one of the lower floors represented a period of time when his work was relatively unaffected by the problems on the remaining higher floors. For the analysis that was performed, the 2nd floor was chosen as the "baseline" when relatively few problems were encountered. Also, the 2nd floor was chosen because it was very similar to the remaining floors going up while the first floor had a different configuration then the remaining upper floors.

Unfortunately, few detailed records of the exact manhours spent per floor were maintained. However, the following data was compiled by the painter for the work on the 2nd floor:

The painter's estimated cost based on the schedule of values/costs for the 2nd floor was:	
Walls (19,000 square feet @ $.45 per square foot)	$8,550
Multi-color (4,000 square feet @ $.70 per square foot)	$2,800
Doors (40 @ $40 each)	$1,600
Mechanical room	$5,950
Punch out	$2,100
TOTAL	$21,000
The painter's actual cost for the 2nd floor (adjusted to remove the cost of change order work):	
Labor	$8,548.43
Materials	$3,560.15
TOTAL LABOR AND MATERIALS	$12,108.98
Overhead @ 15%	$1,816.35
TOTAL ACTUAL COST EXCLUDING PROFIT	$13,925.33

Based on these costs compiled by the painter for the work on the 2nd floor, it appeared that when the painter did not experience the problems he had on the higher floors, he was able to perform the work well within his schedule of value cost. The painters records also indicated that 585 man-hours were expended on the second floor.

The painter, based on his review of the project, selected the 13th floor as one that was representative of the kinds of problems he had encountered on most of the upper floors. In turn, the painter compiled the following information regarding his performance on the 13th floor:

The painter's estimated cost based on the schedule of values/costs for the 13th floor was:	
Walls (18,000 square feet @ $.45 per square foot)	$8,100
Multi-color (4,000 square feet @ $.70 per square foot)	$2,800
Doors (40 @ $40 each)	$1,600
Mechanical room	$5,500
Punch out	$2,000
TOTAL	$20,000
The painter's actual cost (adjusted to remove any change order work)	
Labor	$21,074.69
Materials	$3,661.70
TOTAL LABOR AND MATERIALS	$24,736.39
Overhead @ 15%	$3,710.46
TOTAL ACTUAL COST EXCLUDING PROFIT	$28,446.85

In contrast to the painter's work on the 2nd floor, the actual costs for performing the work far exceeded the schedule of values/costs for the 13th floor.

The comparison of the painter's actual costs to the schedule of values/costs on the 2nd floor confirmed the painter's ability to complete the work within the schedule of values/costs. Accepting this as a verification of the reliability of the schedule of values/costs, the comparison of the schedule of values/costs and the actual costs for the 13th floor supported that the work was performed with a reduction in efficiency. Given the problems as described by the painter on the upper floors, this would be expected.

To calculate the increased costs associated with the inefficiency, the following analysis was performed:

The scheduled value of the painter's activities were as follows:

Basement	$3,000
1st Floor	$14,000
2nd Floor	$21,000

3rd Floor	$20.000
4th Floor	$29,000
5th Floor	$28,000
6th Floor	$22,000
7th Floor	$18,000
8th Floor	$20,000
9th Floor	$20,000
10th Floor	$20,000
11th Floor	$20,000
12th Floor	$20,000
13th Floor	$20,000
14th Floor	$20,000
15th Floor	$20,000
16th Floor	$24,000
17th Floor	$26,000
18th Floor	$4,000
Atrium	$10,000
Garage	$10,000
Misc. Ext.	$10,000
TOTAL PAINTING	$399,000

If painting work was expended in approximately the same proportion for all painting work as for the 2nd floor, then the estimated painting labor expenditure in man-hours would have been 11,115 man-hours ($399,000 × 585/$21,000). According to the painter's records, the actual man-hours expended were 26,677, an overrun of 15,562 man-hours.

The total expended man-hours included time spent for change order work. Though the exact number of man-hours devoted to change order work could not be determined, if the labor component of change order costs was similar to the proportion for the 2nd floor, then an adjustment can be made to account for change orders. The total value of change orders appears to be $112,117. The adjustment would then be 3,123 man-hours (585/$21,000 × $112,117).

Taking the total labor expended by the painter of 26,677 man-hours and subtracting the 3,123 man-hours for change orders, yields a total labor expenditure of 23,554 man-hours. Subtracting the 11,115 man-hours that the painting work should have taken results in a labor overrun caused by the inefficiency of 12,439 man-hours. Applying a burdened labor rate of $12.31/man-hour to these man-hours yields a labor cost attributable to inefficiency of $153,124.09. Adding 15% for overhead yields a total added cost, not including profit, of $176,692.70.

Discussion

Is the preceding approach the preferred method to be used? Probably not. Our preference would be to have detailed records of every man-hour spent on each specific activity. Similarly, we would want a detailed accounting of man-hours spent on all change order work. In that same context, each and every interference on each and every floor or location should have been documented. While the project participants can testify in court or arbitration about what occurred and where, and man-hour records can be produced, hard documentation that is specific is far more persuasive.

5
Regression Analysis

We know, the alarm bells are going off. You may be thinking that the authors have some weird idea that statisticians are reading this book. No worries. We haven't flipped out on you. Other chapters refer to regression analysis and, hopefully, you will recognize how important it is to any claim or argument about loss of productivity. We will simplify this to a level that all can assimilate. And to the statisticians who are reading this, we apologize. Suffice it to say, there are numerous more rigorous writings on the subject of regression analysis. Please feel free to delve further.

In simple terms, regression analysis is a statistical method that allows you to examine the relationship strength between variables. To give that context to this book, let's take the hypothesis that increasing overtime hours (or longer or more workdays) will result in a reduction in productivity. This is one of the more common assertions associated with a claim of lost or reduced productivity. This particular assertion will be addressed in more detail in subsequent chapters. But let's get back to the topic at hand.

When we are speaking of regression analysis, we will be confining our discussion to a simple linear regression analysis. As we said, in regression analysis we are examining and testing the strength of the relationship between two variables. More than two variables can be integrated into the test but that leads you to a much more sophisticated form of regression analysis. Let's stick to just two factors that we want to explore to determine if there is a relationship. In essence, a regression analysis plots a set of data based on the value of the two variables you are testing. From the plot of the data, a linear regression analysis will plot a straight line that "fits" as best as possible to the plot of the data.

Let's start with a simple non-construction example to further illustrate the point of linear regression, and we'll use the apparent relationship between exercise and blood pressure as an example. Now, we recognize that there may be other factors, or variables, in the human body that influence

Ted Trauner, Chris Kay and Brian Furniss. Productivity in Construction Projects, (63–72) © 2022 Scrivener Publishing LLC

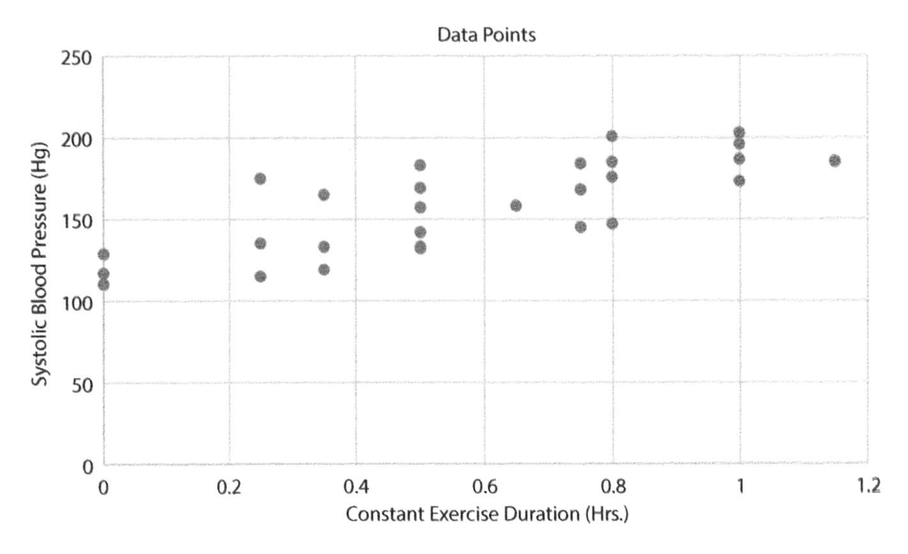

Figure 5.1 Sample comparison of exercise duration and systolic blood pressure.

blood pressure (e.g., genetics, diet, stress, etc.) but with this example we are simply testing the relationship between exercise and blood pressure. Figure 5.1 is an example of a plot of data collected.*

In this example, we are testing the relationship between measured systolic blood pressure and hours of exercise each day. For this example, do hours of exercise (the independent variable or "X") and systolic blood pressure (the dependent variable or "Y") have a strong relationship? Figure 5.2 shows the best fit of a line for the data points acquired.

Figure 5.2 shows a simple plot of data and a simple line showing a best fit for that data. The closer the data points are to the line, the more accurately the line *may* allow us to predict the relationship between the two variables. The reader can see that the best fit line shows the predicted value of systolic blood pressure (Y) for any given value of exercise duration (X) *based on the sample data available*. The more variation in systolic blood pressure (Y), the less that hours of exercise (X) demonstrates that it's a strong predictor of systolic blood pressure.

Although we have offered a simple graph with plotted data points, regression analysis is not a simple exercise. It involves testing a variety of

* Warning: This data is not real. It was completely invented by the authors for illustrative purposes only.

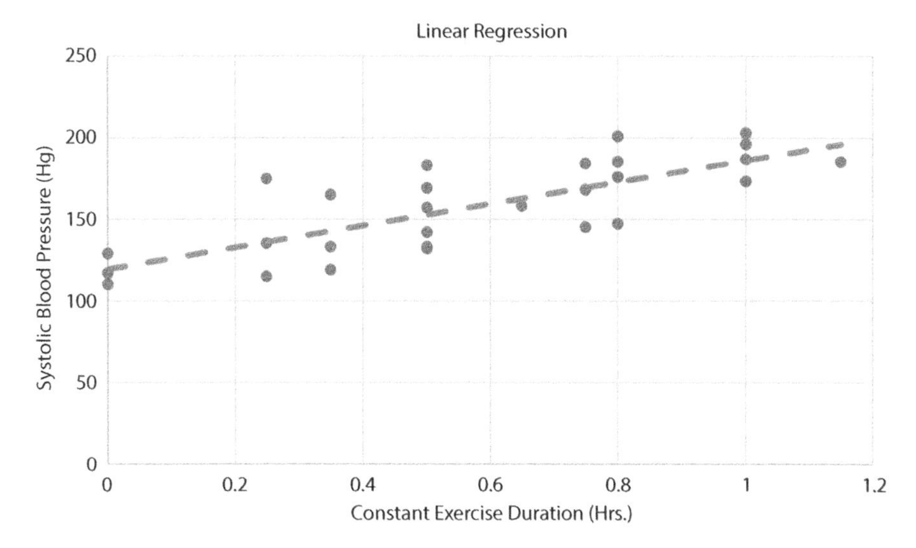

Figure 5.2 Simple linear regression of exercise duration and systolic blood pressure.

variables and then calculating a standard of reliability. Statistics explain how a formula for the line is determined and also how to check the predictability of your model. We won't get into the mathematical formulas, but we think a little more detail is important. In regression analysis simple formulas allow you to calculate a value termed "R." R is generally referred to as the coefficient of correlation. That term can be a bit misleading. Oftentimes, when a regression analysis is run, the "analyst" will plot the data, perform the mathematics, and determine a value for R. If the regression analysis generates a value of R of 0.71, the analyst will then conclude that their model is accurate 71% of the time. But that would not be accurate. A little more digging into statistics tells you that R must be squared (multiplied by itself) to get a value for R-squared (R^2). R-squared, which is also known as the coefficient of determination, is a statistical measure of how close the data are to the fitted regression line. R-squared is always somewhere between 0 and 1, and the closer to 1, the better the strength of the relationship may be. If we take the 0.71 value for R and square it, we get an R squared value of 0.50. Effectively, that means our model shows that the X-variable may be an "accurate" predictor of the Y-variable only 50% of the time. In simple terms, the accuracy of X predicting Y is the same as predicting when the flip of a coin will turn up heads.

Figure 5.3 shows the linear regression line, equation, and R-squared value for the data.

Note that for a simple linear regression, the formula used is a simple slope-intercept equation that most learned in middle school (Y=mX+b). The predicted value of systolic blood pressure (Y) can be calculated by taking the slope of the line (m), multiplying it by the hours of exercise (X), and then adding the intercept value (b = value of Y when X equals zero). This may seem a little like overkill, but we wanted to use this example to illustrate the simplicity of the mathematics. When applied, this equation will allow the analyst to predict the value of Y for any value of X. Now, to determine the strength of that relationship, we need to reference R-squared, which is also shown in Figure 5.3. By now, an analyst has already determined that an R-squared value of 0.59 means that the strength of the relationship between hours of exercise (X) and systolic blood pressure (Y) is not very strong—slightly better than a coin flip. This also tells us that some other variable besides the duration of exercise influences systolic blood pressure. Other variables may be the type of exercise performed, genetics, prior cardiovascular health, diet, etc. The point is that exercise duration (X) does not always predict systolic blood pressure (Y) and the strength of that relationship is not very good.

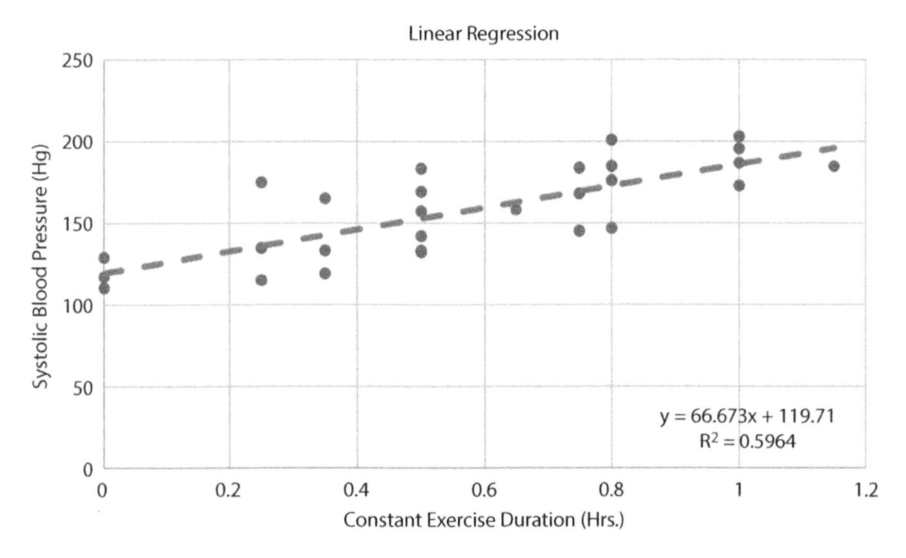

Figure 5.3 Linear regression equation and R-squared.

The reader may have observed over the last several paragraphs on "Regression Analysis" that the authors used the term "may" quite a bit. For example, the "line *may* allow us to predict the relationship" and "our model *may* allow us to predict the relationship," and so on. The reason we did that was not our fear of absolutes. Rather, we wanted to highlight that even if a strong relationship exists via a high R-squared value, it does not mean that one item is the accurate *cause* of the other. Let's take our blood pressure versus exercise example. Presume that someone ran a regression analysis and determined that exercise is related to blood pressure with an R-squared value of 0.85. In other words, it may appear that exercise is an accurate predictor of blood pressure 85% of the time. However, also presume that people who exercise regularly may also consume a better diet, take multi-vitamins, practice meditation, etc. These other variables may influence or alter the strength of the relationship between exercise and blood pressure because they may influence the exercise variable. One may want to isolate these variables from exercise to accurately determine whether exercise is the only factor causing the 0.85 R-squared relationship value to blood pressure. The old saying "correlation does not equal causation" is true and, perhaps more accurately for this example, "R-squared (determination) does not equal causation."

In cases where there are multiple variables in play, and a lot of dollars at stake, parties will utilize regression analysis to support their case. The important point is that the parties should check all the references concerning predictions for loss of productivity to see if the analysts authoring those studies collected data randomly, if there were other factors that could have influenced the outcome of their analysis, and if the analysts then verified their predictions by some form of a statistical analysis.

The value of a regression analysis is to be able to explain an outcome based on the relationship between certain variables, such as blood pressure and exercise. Now that the concept of regression analysis has been introduced, Case Study #8 provides an example of how regression analysis can be used to determine the most significant variable that affected another variable despite what a "common sense" approach might tell us.

Another point that the reader should appreciate from this case study is that many variables were reviewed and tested before a reliable correlation was determined.

Case Study #8: The Transmission Line

Several years ago, the authors were engaged on a project that had a significant claim concerning loss of efficiency. The project was the construction of a 500KV transmission line. The following information summarizes the key elements of the project and the claim.

The work was located in the northwestern United States in an extremely mountainous area.

Because of the location of the project, access was very limited. Most material equipment and manpower had to be flown in by helicopter.

The work involved the construction/erection of transmission line towers. There were seven different types of towers, including both guyed and self-supporting towers.

The owner supplied the materials for the project. The most significant material was the steel that comprised the structural members for all of the towers. The owner contracted a supplier/materialman to provide the steel based on specifications drawn up by the owner's designer.

The tower steel sections were bolted together in the field with high strength steel bolts and, where necessary, gusset plates.

During the course of erection, the erection contractor complained vigorously of mis-fabricated steel pieces. This mis-fabrication necessitated that the erection contractor ream or redrill thousands of bolt holes in order to make the towers fit together. In fact, the claim asserted that approximately 50,000 bolt holes were mis-fabricated.

As a consequence of the mis-fabricated steel, the erection contractor claimed that his crews suffered tremendous losses in productivity. The claim that was submitted totaled several million dollars.

When we began our analysis of the claim, we first investigated if the steel was in fact mis-fabricated, as claimed by the erection contractor. The results of that review verified that the steel did have significant mis-fabrication problems that necessitated that the erection contractor ream or redrill bolt holes. It must be noted, however, that the count of the number of mis-fabricated bolt holes overstated what really was happening. For example, if three steel pieces came together at a connection and one hole was misaligned, the reaming was done on that one hole in a single operation. The erection contractor counted this as three mis-fabrications. Obviously, this was somewhat misleading.

This project was an excellent example of the benefit of good recordkeeping. The engineering firm that inspected the work kept detailed records for every tower that was constructed. These records included the following information:

Tower type
Number of pieces
Shake out hours
Assembly hours
Crew size
Crew chief/foreman and crew members by name
Time spent for reaming, bolting, etc.
Travel time to assembly sites

Clearly, one could identify the direct time spent for the reaming and redrilling since the records contained that detail. Therefore, an amount of compensation for this extra work was directly calculated. This amount did not even approach the sums claimed by the erection contractor. Therefore, an analysis was required to determine what potential productivity losses might have been experienced because of the extra work that resulted from the mis-fabrications. That analysis was performed in the following fashion.

The overall productivity in man-hours per tower was calculated and plotted over the life of the project. The actual number of holes that had to be reworked was indicated on each individual tower. One would expect that productivity would decrease as the number of mis-fabrications increased on any given tower type. Surprisingly, this was not the case. Had this been so, it would have supported the erection contractor's claim that its losses were a direct result of the mis-fabrication of the steel. It also would have allowed the analyst to develop a measured mile between a least impacted tower and the remaining towers by tower type. Since there was no direct correlation between the hours required to erect a tower and the number of mis-fabricated holes, this could not be done. But this lack of correlation raised a much more significant question. If the time to erect a tower was not directly influenced by the number of mis-fabrications, what factors were affecting the demonstrated productivity? This question led to the second phase of the analysis.

In the continuation of the analysis, our staff looked more closely at what the available information showed to see if there were any obvious elements that could be readily identified that might affect the productivity.

The first factor that was recognized was that the productivity increased significantly about a third of the way into the project. This increase in productivity corresponded directly with a change by the erection contractor in the overall project management team. This initial clue led to the development of a list of all possible factors that might have affected the productivity. Some of these factors included steel-related items such as number of pieces, number of mis-fabrications, etc., and non-steel-related items such as tower location, travel time, crew chief/foreman, and crew staff.

With the various factors identified, a computer model was constructed to analyze the data to determine the correlation of each possible variable with the productivity demonstrated. The results of the regression analysis were far different than what was surmised when the analysis began.

When compared on a tower-type basis, the most dominant factor effecting productivity was the crew chief/foreman. Also, the variation in productivity among crew chiefs was dramatic. The "good foreman" always had good productivity regardless of the tower type and regardless of the number of mis-fabrications on a tower. In fact, there was little difference in the productivity by tower type for a good foreman even if one tower had two or three times more mis-fabrications than other towers in the group. In other words, a good foreman got good results whether the tower had mis-fabrications or not.

The ultimate result of the analysis was that it could be shown with a reasonable level of confidence that the bulk of the productivity variations were not due to the steel problems. Instead, they were directly related to the management of the crew by the crew chief.

An analysis of this type was possible only because of the nature of the work, the simplicity of the alleged problem that was claimed, and the large volume of specific detailed information for the entire tower population. This volume of data allowed for an analysis of the entire population and did not require that a model be based on a limited population and extrapolated or inferred to the rest of the population.

The lesson learned from this case study is simple. While "common sense" would lead one to believe that the amount of mis-fabrications would control the time spent on erection, something more is needed to support that supposition. Hence, a regression analysis. And in this instance, it was several regression analyses that finally demonstrated a significant correlation between increased man-hours of effort and the specific foreman running the tower crew.

6
Learning and Experience Curves

Sometimes not enough data is available to do a straightforward measured mile. Similarly, there are times when an analyst recognizes that an interruption or disruption on a project may adversely affect the learning process that a work crew develops as they work through a structure. It is on these occasions that one might consider learning curve effects. Let's start with the basics.

Oftentimes the terms learning curve and experience curve are used interchangeably. Some writings on the subject do offer some difference between the two. In simple terms, experience is gained with every job and task a workman performs in his particular field. For example, an electrician with 10 years of experience will have a different skill level than an electrician with one year of experience. Consequently, the more "experienced" electrician may well be able to perform a task more quickly or more efficiently. The 10 years of experience have allowed the electrician to develop his/her own techniques and skills such that most tasks he/she will perform can be performed in less time. A learning curve, on the other hand, is project specific. In other words, once a craftsman becomes familiar with the project, the location of materials, the supporting staff, the location of tools, etc., he/she should be able to perform repetitive tasks more rapidly (efficiently) as they are repeated throughout the project. So, as we view these two terms, experience curve and learning curve, the experience curve is broader and flows from the aggregate experience that someone has over many different projects. The learning curve is narrower and relates to what is "learned" on a specific project.

The concept of learning curves is not new. It was first noted in the 19th century but wasn't quantified until 1936 during the production of airplanes at Wright-Patterson Air Force Base. The concept in simple terms holds that for every doubling of a task, there will be some reduction in the time to perform the task. Every task has its own unique learning curve. According to Wikipedia, the Boston Research Group (BRG) observed experience

Ted Trauner, Chris Kay and Brian Furniss. Productivity in Construction Projects, (73–76) © 2022 Scrivener Publishing LLC

curve and learning curve effects in the 1970s. Their research noted learning curve effects for different industries ranging from 10 percent to 25 percent. What those numbers mean is simple. A 90% learning curve means that the cost of a doubling of output would be reduced by 10%.

Can the concept of a learning curve be applied to construction? Yes, it can. We have included the case study shown below that demonstrates this. The main point that should be recognized is that any benefits of a learning curve may well be eliminated because of interruptions in the performance of the work. Therefore, in the situations where not enough data exists to perform a reasonably good measured mile, one alternative may be to measure any alleged loss of productivity by calculating a demonstrated learning curve and then applying that to the period of work where "changes" negated the learning curve benefits.

The following case study exemplifies the use of learning curves.

Case Study #9: Elevated Rail Project

This case study is a bit different since it intertwines the measured mile analysis with a learning curve.

The project involved the reconstruction of portions of an elevated rail line. The project had four separate contracts, all with the same contractor. During the course of the projects, the Occupational Safety and Health Administration (OSHA) made a change to its regulations and imposed new requirements for workers exposed to lead. The changed regulations were not amended until after the contracts were bid on, awarded, and under construction. The four contracts totaled approximately $89,000,000. The claims for these four projects totaled approximately $7,000,000. According to the contractor, the productivity losses took three forms:

1. A significant amount of time was consumed in conforming to the new requirements of the regulations. These included lost time for training, medical exams and tests, and similar activities. The direct costs for this time were compiled by the contractor from project records.
2. The second type of loss involved the time required for donning and removing respirators, Tyvek and Nomex suits, and other lead-related safety equipment before lead work was performed, for breaks and lunch, and at the end of the day.

Also included was cleanup time during breaks and lunch and showers at the end of the day. The contractor determined the time lost for these activities by timing the actual suit up and cleanup activities of its crews. This data was then used to calculate an average time lost per shift for these activities.

3. The last area of reduced productivity was the time lost during performance of the work, related primarily to the inconvenience associated with wearing full head respirators and Tyvek or Nomex suits.

Obviously, the first two preceding items could be quantified by direct measurement of the time involved. The third element, however, was more challenging. In order to perform a measured mile to support an amount claimed for item three, a demonstrated period of work prior to the regulation change was necessary, along with detailed records of the time spent on common tasks. One of the projects met this requirement and was used as a baseline for measurement. Some of the columns in the bents were replaced before the OSHA change and some afterwards. For the actual analysis, the column bents were grouped, and each group of columns was numbered. For each group of bents, man-hours were tabulated for each bent from the date the truss load was transferred to the cribbing to the date the load was returned to the reconstructed bent. Only hours specifically related to bent reconstruction work were included. Based on the detailed project records, the initial bents required 330 man-hours per bent and there was an initial improvement down to 318 man-hours per bent. This productivity then decreased to 366 man-hours per bent and over time the productivity improved to 305 man-hours per bent. Clearly, one would not expect an improvement in the productivity of the bent construction given the nature of the OSHA requirements. However, if one reflects on it, one should have expected that bent construction over the life of the project should have improved with each group of bents. The expectation that repetitive work will become more productive as more repetitions are performed is often called the "learning curve" effect. A learning curve is a plot of the cumulative average man-hours required to perform a task as the number of repetitions increases. The basic observed relationship is that for each doubling of units produced, productivity improves by a constant percentage. For example, for a 90% learning curve, the cumulative average manpower required to perform each doubling of units will be 90% of the previous doubling. Different types of work have different learning curves. However, given the nature of this project and the OSHA regulation change, it was not possible to determine the exact learning curve that would be applicable.

It was possible, however, to plot a learning curve for the precast deck installation work. Since the precast deck installation work was not affected by the OSHA change, an unimpacted learning curve could be plotted for this work. This curve showed that after each doubling the learning curve was 79%. Though there is a difference in the work, a second learning curve was plotted for the stringer replacement work. The results of this portion of the analysis were that the learning curve factor for the stringers was 73.28%.

Without going into all of the charts and plots for the work, let's cut to the chase. The productivity for the columns prior to the new OSHA requirements was 323.63 man-hours/column. This represents the contractors unimpacted productivity prior to the revised lead laws. The average productivity achieved on the remaining work was 342.78 man-hours/column.

To calculate the loss of efficiency for the contractors work there are two possible calculations. One with the learning curve factor of 79% and the other with the learning curve factor of 73%. The more conservative choice is the 79% one. Therefore, the calculated loss of efficiency experienced by the contractor because of the revised OSHA requirements would be:

$$323.63 \text{ man-hours/column} \times 79\% = 255.67 \text{ man-hours/colum}$$

The lost productivity is then calculated as:

$$(342.78 - 255.67) \div 255.67 \times 100\% = 34.07\%$$

Discussion

This is an interesting case study. It is interesting because while we can determine a man-hour amount for lost productivity by just a measured mile as we did initially, we need to keep our thinking caps on and recognize that repetitive work normally yields a long-term benefit in terms of the efficiency of performing the operation. Not very far afield from the manufacturing facility concept.

7

The Kitchen Sink Approach – Blending Methods Together

There are some projects that experience problems that may require some combinations of approaches in order to quantify as best as possible the increased costs that occurred. Other factors besides just a measured mile may enter into the mix. Consequently, the analyst must be aware of all the factors that occurred and put on their thinking caps in order to persuasively incorporate and quantify those factors. The following case study is one such project and it is included to demonstrate the thoroughness and inclusiveness that must be incorporated into any presentation.

Case Study #10: The High-Rise Hotel

During the review and analysis of a major dispute on a high-rise hotel, several factors were evident that adversely affected the productivity of the workforce. This example is a good illustration of how to approach an inefficiency situation when you have more than one factor affecting productivity and when you do not have perfect or totally complete information.

The hotel was 55 stories, and because of a number of factors, was forced into an accelerated construction schedule. Besides numerous and significant delays, there was a tremendous number of design changes and clarifications. As a consequence of this, there was a significant amount of change order work, most of it being performed on a time and material basis. Compounding the problem of delays and changes was the dilemma of inadequate vertical transportation. In some instances, lifts ran only to the 40th floor and workers had to climb ladders for the remaining 15 floors. The project was accelerated to an extreme. To put that into perspective, the electrician, about whom we will be discussing, was forced to increase

Ted Trauner, Chris Kay and Brian Furniss. Productivity in Construction Projects, (77–96) © 2022 Scrivener Publishing LLC

the manpower level to 650 tradesmen. The original staffing was at the 150 tradesmen level.

In this project, the productivity of the tradesmen was affected by overtime, crowding, significant numbers of disruptive changes, and increased manpower levels. Simultaneously, the vertical transportation problem was causing wasted man-hours throughout the day. In order to analyze the project as accurately as possible, the effects of vertical transportation were isolated to allow accountability for those man-hours and to exclude them from any other quantification of loss of efficiency. Therefore, we will first describe how the vertical transportation problem was addressed and then proceed to the measured mile for the other factors that affected productivity.

Vertical Transportation

Vertical transportation was crucial to the productivity of the prime contractors on the project. Time spent waiting for and riding the vertical lifts was not productive to the respective contractors. Thus, the longer the wait, the greater the loss of productive man-hours.

Numerous complaints were submitted by the prime contractors to the construction manager concerning the lack of adequate vertical transportation. At the midpoint of the project, agreements were entered into, compensating the contractors for problems to that point and promising to solve the problem by increasing the available vertical transportation. Along with this, the contractors were required to increase their respective manpower levels to meet the accelerated schedule.

After the agreement, however, the vertical transportation did not increase. Prior to the agreement, four hoist cars were available with a capacity of 25 personnel each or a total of 100 personnel. Shortly after the agreement, one hoist of two cars was dismantled and three elevators were placed into operation. The elevator cars, however, could only accommodate 15 personnel each. Therefore, the vertical transportation decreased from 100 personnel to 95 personnel. Therefore, instead of an increase as promised, the lift capacity was reduced.

In order to analyze the man-hour losses from inadequate vertical transportation, queuing theory was utilized. Queuing theory is a statistical method of analyzing waiting lines, such as at a bank, a fast-food restaurant, or in this case, at an elevator or a hoist. Queuing theory was used to determine the average daily time spent by each worker waiting on vertical transportation. The detailed analysis using queuing theory will not be reproduced herein. It should be noted that when multiple factors affect the productivity, the analyst must sort out these factors and apply whatever analytical techniques are necessary in order to attain accurate results.

The following summary provides an overview of the general approach to the problems with the vertical lifts and the other productivity losses.

Based on project information, the average electrician made six trips on vertical lifts each day.

Up to the work area at the start of the day.
Down to the street level at the end of the day.
Down to the street level for lunch.
Return to the work area from lunch.
Down to the restroom.
Return to work area from the restroom.

The restrooms were on the ninth floor. As a result, vertical transportation was required to go to the restroom. The six trips were independent of each other. Thus, the waiting time for each trip was analyzed separately. The analysis was also divided into three time periods to account for changes in the number of lifts available to construction personnel and the number of personnel on the job.

First Time Period

Trip to work area at the start of the day.
Four personnel hoists were available to construction personnel.
Number of lifts = 4.

A detailed analysis of the electrical contractor's timesheets was conducted to determine where personnel were working throughout the hotel. For the purpose of this analysis, the floors of the hotel were divided into groups. The subforemen's timesheets were used to determine the number of men working in each group of floors on a daily basis. Three separate weeks were used in this analysis. The total number of men per group of floors was calculated for all three weeks. These totals were used to calculate the percentage of men per group of floors.

Men working from the sub-basement to the second floor did not require vertical transportation. Approximately 71% of the electrical contractor's personnel required vertical transportation. The time for the hoists to reach each group of floors was estimated based on discussions with the electrical contractor's project manager and the site engineer. The estimated times were multiplied by the number of men per floor to determine cumulative service time per floor.

The average service time was the average time workers actually spent on the lifts during a trip. Based on the project information, the average service time was approximately eight minutes.

Applying queuing theory to this first time period resulted in an average waiting time of approximately 54 minutes.

Return to street level at the end of the day.

The number of lifts and service rate remained the same. Due to overtime, all trades did not leave at the same time. According to the electrical contractor's project manager and the site engineer, all personnel left within two hours. Again, using queuing theory, the average waiting time was approximately 9 minutes.

Go down to the street level for lunch.

The number of lifts and service rate remained the same. All personnel did not go down to the street level for lunch. According to the electrical contractor's project manager, 85% of the personnel went to the street level for lunch and all personnel took lunch within one hour. For the lunch trip the average waiting time was approximately 15 minutes.

Return to the work area after lunch.

This trip was the same as the previous trip going to lunch. Thus, the average waiting time was 15 minutes.

Trip to the restroom.

The number of lifts and service rate remain the same. As previously stated, the average worker made one trip to the restroom each day. The trips were assumed to occur within a five-hour period of the day, excluding lunch and the start and end of the day. The average waiting time was approximately 8 minutes.

Return to work areas from the restroom.

This trip was the same as the previous trip going to the restroom. Thus, the average waiting time was 8 minutes.

The total average waiting time for all six trips was calculated as follows:

Trip	Average waiting time (Minutes)
1	53.62
2	8.58
3	15.02
4	15.02
5	8.02
6	8.02
TOTAL	108.28 Minutes

Thus, the average loss of productive man-hours per worker for the time period analyzed, was 108.28 man-minutes per day or 1.80 man-hours per day.

Second Time Period

During this period, two outside personnel hoists and three elevator cars were available to the construction personnel for a total lift capacity of 95 men.

Using the same approach as the first time period shows that the total average waiting time for all six trips was calculated as follows:

Trip	Average waiting time (Minutes)
1	163.64
2	8.42
3	13.75
4	13.75
5	8.00
6	8.00
TOTAL	215.56 Minutes

The average loss of productive time for the second time period analyzed was 215.56 man-minutes per day or 3.59 man-hours per day.

Third Time Period

During this period vertical transportation available to construction personnel varied due to the operations of the hotel and removal of the hoists. Due to this variation in available vertical transportation, a determination of the average capacity of the lifts during this time period was not feasible. The capacity of vertical transportation during this time period did not surpass the capacity in the previous time period. Therefore, the average capacity of the lifts during the previous time period will be used to calculate losses during this time period. From project daily reports, the average number of personnel on the job was 100.

The calculations for this period were similar to the calculations in the previous two time periods. The results are summarized as follows:

Trip	Average waiting time (Minutes)
1	8.00
2	9.00
3	8.00
4	8.00
5	8.00
6	8.00
TOTAL	48.00 Minutes

The average loss of productive time for this third time period was 48.00 man-minutes or 0.80 man-hours.

The total man-hours lost due to vertical transportation are shown in Table 7.1.

Since not all of the electrical contractor's workforce required the use of vertical transportation, the actual man-hours lost by the electricians due to inadequate vertical transportation was calculated as follows:

Man-Hours Lost	-	101,330.45 MH
Percent of Electrical Contractor's Workforce Requiring Vertical Transportation	-	71.2%
Actual Man-Hours Lost (101,330.45 × 71.2%)	=	72,147.28 MH

Table 7.1 Man-hours lost due to vertical transportation.

Time period	Month	Contractor average manpower	Workdays	Man-hours lost per day	Total man-hours lost
1	March	129	5	1.80	1,161.00
	April	269	17	1.80	8,231.40
2	April	269	5	3.59	4,828.55
	May	286	22	3.59	22,588.28
	June	277	20	3.59	19,888.60
	July	231	22	3.59	18,244.38
	August	206	22	3.59	16,269.88
	September	161	4	3.59	2,311.96
3	September	161	16	0.80	2,060.80
	October	110	23	0.80	2,024.00
	November	60	19	0.80	912.00
	December	51	20	0.80	816.00
	January	46	22	0.80	809.60
	February	33	20	0.80	528.00
	March	41	20	0.80	656.00
				TOTAL:	101,330.45

Expected Losses

Even in an ideal situation, where the electrical contractor had sole use of the lifts, a loss of productive man-hours would have occurred. Consequently, some losses of productive man-hours due to vertical transportation should have been expected.

As with the actual losses, the expected losses were calculated using queuing theory. The calculations were based on the electrical contractor having sole use of the lifts.

Based on the project information and queuing theory, the electrical contractor's expected loss per day for each time period was 48 minutes, or 0.8 man-hours.

Using this expected daily loss, the electrical subcontractor's total expected loss of man-hours due to vertical transportation was:

Expected Lost Man-Hours	-	30,728.80 MH
Percent of Electrical Contractor's Workforce Requiring Vertical Transportation	-	71.2%
Actual Expected Lost Man-Hours (30,728 MH × 71.2%)	=	21,878.90 MH

Unexpected Losses

The difference between the electrical contractor's actual lost man-hours and expected lost man-hours was the unexpected lost man-hours incurred by the electrical contractor due to inadequate vertical transportation. The unexpected loss was calculated as follows:

Actual Lost Man-Hours	-	72,147.28 MH
Expected Lost Man-Hours	-	21,878.90 MH
Unexpected Lost Manhours	=	50,268.38 MH

Additional Lost Man-Hours

The calculation of lost man-hours using queuing theory was based on the assumption that all vertical lifts operated nonstop during the workday. This was not the case. Mechanical breakdowns, lack of operators, maintenance, and other uses for the lifts resulted in unavailability of lifts to construction personnel. The unavailability of lifts caused the electrical contractor to incur additional lost man-hours. These additional lost man-hours were not accounted for in the queuing theory calculations and were thus calculated separately.

The electrical contractor's general foreman maintained a log for five months of the project. The log noted that delays due to lift unavailability occurred on 38 out of 58 logged days, or 65.5% of the time. The number of lifts unavailable and the duration of the unavailability varied. A review of the log indicated that the delays were generally greater than one hour. Thus, the average delay was at least one hour. The total number of man-days during the period of the problems was 38,411.

Additional lost man-hours were calculated as follows:

Percentage of Days with Delays	-	65.5%
Total Man-Days with Delays (38,411 Man-Days × 65.6%)	=	25,159.20 Man-Days
Percentage of Electrical Contractor's Workforce Requiring Vertical Transportation	-	71.2%
Actual Electrical Contractor Man-Days with Delays (25,159.20 Man-Days × 71.2%)	=	17,913.35 Man-Days
Lost Man-Hours Due to Delays (17,913.35 Man-Days × 1 hr/day)	=	17,913.35 MH

Thus, the electrical contractor lost an additional 17,913.35 MH due to unavailability of elevators to construction personnel.

Total Lost Man-Hours Due to Inadequate Vertical Transportation

The electrical contractor's total lost man-hours due to inadequate vertical transportation was calculated as follows:

Unexpected Lost Man-Hours Waiting for Operational Lift	-	50,268.38 MH
Lost Manhours Due to Lift Unavailability	-	17,913.35 MH
Total Man-Hours Lost by the Electrical Contractor Due to Inadequate Vertical Transportation (50,268 MH + 17,913.35 MH)	=	68,181.73 MH

Productivity Losses

The electrical contractor also incurred significant increased costs due to productivity losses. The losses of productivity were a direct result of delays and disruptions caused by:

- Change orders
- Design conflicts
- Acceleration of the schedule
- Inadequate vertical transportation

The delays and disruptions associated with the change orders were created by:

- The lengthy change order approval process
- The untimely introduction of changes
- The effect on the electrical contractor's work by changes to other subcontractor's work
- Holds on various portions of the building
- Changes to changes by the owner

The planned flow of the building construction was severely and adversely altered due to these changes. This out-of-sequence work caused the electrical subcontractor to incur additional labor costs as a result of having to perform numerous additional tasks that were not contemplated at the time of the agreement. Some of these tasks were:

- Additional man-hours were expended to secure materials that were brought to the work area but could not be installed.
- Additional man-hours were expended as a result of having to return to a work area that could not be completed in sequence with other work. The specific additional tasks required to return to a work area included:
- Review drawings and specifications to determine what and how to perform the work.
- Determine what tools, equipment, and material are required to perform the work. Obtain those required materials.
- Determine what materials, if any, had already been stored near the work area.

- Coordinate the electrical work with other trades working in the area.
- Determine what materials had already been installed.
- Demobilize the craftsmen from other tasks in order to finish the now available work which had previously been left out.
- Perform M/D directed T&M work and return to original contract and lump sum (LS) change order work.

The disruptions encountered resulted from both changes issued to the electrical contractor as well as changes issued to other subcontractors. The electrical subcontractor was compensated for the direct costs associated with the change, but not for all the additional costs both for the change and the base contract work that were incurred as a result of having to perform work out-of-sequence. The resulting loss of productivity was such that it could not be related to a limited number of specific changes.

The design conflicts caused delays and disruptions due to an inadequate request for information (RFI) system, lack of engineering on the project site, and inadequate control of coordination among subcontractors by the construction manager. The design conflicts and poor coordination affected the electrical contractor's work in the same manner as the changes.

The acceleration of the schedule resulted in stacking of the trades and increased manpower. Thus, areas of the hotel were overcrowded with tradesmen. The overcrowding caused further losses in productivity.

Vertical transportation was a problem on the project. Inadequate vertical transportation has already been discussed.

The losses caused by change orders, design conflicts, acceleration of the schedule, and inadequate vertical transportation can be demonstrated and documented. The following narrative demonstrates the productivity losses using a measured mile approach.

In general, the measured mile approach determines lost productivity by comparing least impacted work with impacted work. The difference between the impacted and least impacted productivity equals the loss in productivity for the activity. From this difference, the percent of inefficiency is calculated.

It must be noted that the loss of productivity was calculated based on the electrical contractor's actual demonstrated performance on the project. It was neither based on industry standards nor on estimates. It was a detailed measure of the man-hours expended in excess of what was demonstrated on baseline work on the job.

The loss of productivity was measured based on the available documentation. Of the various activities performed by the electrical contractor, accurate records existed for guest level rough-in and electrical finish work. Rough-in work consisted of floor layout, installing conduit, and installing back boxes. Back boxes were industry approved, steel rectangular boxes, closed in the back and open in the front, in which are mounted electrical devices and to which conduit is connected. Electrical finish work consisted of installing fixtures, devices (switches and receptacles), and device cover plates in the guest rooms. The data available did not facilitate analysis of the public areas.

In order to best demonstrate the impact of the changes on the electrical work, a comparison was made on the actual documented productivity of the rough-in and finish work. It is reasonable that rough-in and finish work would exhibit the impacts of inefficiencies due to change orders, design conflicts, acceleration of the schedule, and inadequate vertical transportation.

Data Assembly and Calculations

The electrical contractor's subforemen's timesheets were used to determine the total number of man-hours spent per floor on rough-in and finish work. Due to the nature of the data, the floors were broken down into groups. Data was not available for rough-in on floors 10 through 14 and finish work on floors 20 and 41.

The number of rooms per floor was determined from the project drawings. Some floors had non-standard rooms which were larger than the standard guest room. In order to accurately compare the rough-in and finish work using man-hours per room, the non-standard rooms were considered to be more than one room depending on their size and the amount of electrical work involved. Following is a summary of the non-standard rooms and their standard room equivalency:

Non-standard room	Standard room equivalency
Parlor	1
Hospitality Suite	2

Luxury Suite	2
Bedsitting Suite	2
Concierge Lounge (28th Floor)	2
Concierge Lounge (29th Floor)	8
Vice Presidential Suite (43rd Floor)	7
Vice Presidential Suite (44th Floor)	10
Presidential Suite (42nd Floor)	16
Presidential Suite (44th Floor)	12

The productivity was calculated by dividing the number of man-hours expended per activity by the number of rooms per floor or group of floors. For example:

ROUGH IN:

Floors 15 to 21

Total Number of Rooms	-	400 Rooms
Total Labor Spent on Rough-In	-	2,353.5 MH
Demonstrated Productivity	=	5.9 MH/Room
(2.353 *MH*	÷	400 *Rooms*)

FINISH WORK:

Floors 12 to 15

Total Number of Rooms	-	230 Rooms
Total Labor Spent on Finish Work	-	687.0 MH
Demonstrated Productivity	=	2.99 MH/Room
(687.0 MH	÷	230 *Rooms*)

Based on similar calculations, the results in Table 7.2 and Table 7.3 were obtained for guest-level rough-in and finish work:

Table 7.2 Guest level rough-in.

Floor	Number of rooms	Man-hours expected	Productivity (MH per room)
10 - 14	Insufficient Data		
15 - 21	400	2,353.5	5.9
22 - 27	340	2,335	6.9
28 - 31	240	2,039	8.5
33 - 35	180	1,796.5	10.0
36 - 37	100	1,299.5	13.0
38 - 41	240	1,666.5	6.9
42 - 44 & 32	248	4,579	18.5

Table 7.3 Guest-level finish work.

Floor	Number of rooms	Man-hours expected	Productivity (MH per room)
10	60	198.0	3.30
12–15	230	687.0	2.99
11, 16–18	230	752.5	3.27
19, 21–26	400	1,936.0	4.84
27–31	300	1,291.0	4.3
33–36	230	935.5	4.07
37–40, 32	290	972.5	3.35
41–44	184	1,075.0	5.84

As previously discussed, extra work due to change orders caused the electrical contractor to divert effort from base contract work. The following Tables 7.4. and 7.5 illustrate the amount of T&M work completed during the completion of rough-in and finish work.

Table 7.4 Summary of T&M work during performance of rough-in.

Floors	Dates of rough-in completion	Number of workdays	Man-hours of T&M work performed	T&M man-hours performed per day
15–21	08/28/84 - 11/01/84	57	1,601.5	28.1
22–27	10/26/84 - 12/10/84	31	902.5	29.1
28–31	11/26/84 - 01/25/85	42	6,138.0	146.1
33–35	01/14/85 - 03/03/85	35	4,926.0	140.7
36–37	03/04/85 - 04/03/85	23	3,124.5	135.8
38–41	03/29/85 - 05/01/85	23	2,567.5	111.6
42–44 & 32	04/18/85 - 07/10/85	60	20,222.0	337.0

Rough-In

As shown in Table 7.4, the lowest rate of T&M change order work occurred during the completion of rough-in on floors 15 to 21. Rough-in on floors 15 to 21 was also completed prior to acceleration of the schedule. Thus, floors 15 to 21 were used as a baseline for the rough-in work. This does not mean these floors were not impacted. Rather, these floors were the least impacted. The measured productivity for floors 15 to 21 was 5.9 MH/room as calculated earlier in this report.

However, this productivity includes inefficiency due to vertical transportation. During the completion of floors 15 to 21, the electrical contractor's workers lost 3.2 man-hours per day due to vertical transportation. Based on this loss, the least impacted productivity was calculated as follows:

FLOORS 15 TO 21:

Total Man–Hours Expended	-	2,353.5 HM
Average Workday	-	8 MH/Day
Total Man–Days Expended (2,343.5 *MH* ÷ 8 *MH/Day*)	=	294.2 Man-Days
Loss Due to Vertical Transportation (294.2 Man–Days × 3.2 *MH/Day*)	=	941.4 MH
Man-Hours Expended Less Man-Hours Lost to Vertical Transportation (2,353 *MH*–941.4 *MH*)	=	1,421.1 MH
Least Impacted Productivity (1,412.1 *MH* ÷ 400 *Rooms*)	=	3.5 MH/Room

Table 7.5 Summary of T&M work during performance of finish work.

Floors	Dates of finish work completion	Number of workdays	Man-hours of T&M work performed	T&M manhours performed per day
10	02/12/85 - 03/30/85	34	3,678.0	108.2
12–15	02/12/85 - 03/05/85	16	1,171.5	73.2
11, 16–18	03/05/85 - 04/04/85	23	3,124.5	135.8
19, 21–26	04/05/85 - 05/08/85	24	3,435.0	143.1
27–31	05/10/85 - 06/06/85	20	6,574.0	328.7
33–36	06/07/85 - 06/28/85	16	6,137.0	383.6
37–40, 32	06/27/85 - 07/26/85	22	9,152.5	416.0
41–44	08/05/85 - 08/22/85	14	7,114.5	508.2

Given the least impacted productivity, the impact on specific floors can be shown as:

FLOORS 38 TO 41:

Total Number of Rooms	-	240 Rooms
240 Rooms × 3.5 MH/Room	=	840 MH Based on Least Impacted Productivity
Actual Man-Hours Spent	-	1,666.5 MH
Extra Man-Hours Expended Due to Inefficiency (1,666.5 *MH* – 840 *MH*)	=	826.5 MH

In other words, on floors 38 through 41, the electrical contractor expended an additional 826.5 MH on rough-in work due to the delays and disruptions previously discussed. Based on similar calculations, the man-hours in Table 7.6 were calculated.

Finish Work

As shown in Table 7.5, the lowest rate of T&M change order work occurred during the completion of finish work on floors 12 to 15. Finish work on floors 12 to 15 was also completed prior to acceleration of the schedule. Thus, floors 12 to 15 were used as a baseline. The measured productivity for floors 12 to 15 was 2.99 MH/room as calculated previously. However, this productivity contains inefficiency due to vertical transportation. During the completion of floors 15 to 21, the electrical contractor's workers lost 2.0 man-hours per day due to vertical transportation. Based on this loss, the least impacted productivity was calculated as follows:

Table 7.6 Rough-in man-hour summary.

Floors	Total man-hours	Least impacted productivity man-hours	Extra man-hours
38–41	1,666.5	840.0	826.5
42–44, 32	4,579.0	868.0	3,711.0
TOTAL:	6,245.5	1,708.0	4,537.5

FLOORS 12 TO 15:

Total Man-Hours Expended	-	687.0 MH
Average Workday	-	8 MH/Day
Total Man-days Expended (687.0 *MH* ÷ 8*MH/Day*)	=	85.9 Man-Days
Loss Due to Vertical Transportation (85.9 Man-Days × 2.0 *MH/Day*)	=	171.8 MH
Man-Hours Expended Less Man-Hours Lost to Vertical Transportation (687.0 *MH*–171.8 *MH*)	=	515.2 MH
Least Impacted Productivity (515.2 *MH* ÷ 230 *Rooms*)	=	2.2 MH/Room

Given the least impacted productivity, the impact on specific floors can be shown as:

FLOORS 27 TO 31:

Total Number of Rooms	-	300 Rooms
300 Rooms × 2.2 MH/Room	=	660.0 MH Based on Least Impacted Productivity
Actual Man-Hours Spent	-	1,291.0 MH
Extra Man-Hours Expended Due to Inefficiency (1,291.0 *MH* – 666.0 *MH*)	=	631.0 MH

In other words, on floors 27 to 31, the electrical contractor expended an additional 631.0 MH on finish work due to delays and disruptions previously discussed. Based on similar calculations, the additional man-hours in Table 7.7 were calculated.

Percent Loss of Productivity

The percent loss of productivity was calculated as follows:

$$\% \, Loss \quad \frac{Rough-In\ Extra\ Man-hours + Finish\ Work\ Extra\ Man-hours}{Rough-In\ Total\ Man-hours + Finish\ Work\ Total\ Man-hours}$$

$$\% \, Loss \quad - \quad \frac{4{,}537.5\ MH + 3{,}121.2\ MH}{6{,}245.5\ MH + 6{,}210\ MH} = 61.5\%$$

Table 7.7 Finish work man-hours summary.

Floor	Total man-hours	Least impacted productivity man-hours	Extra man-hours
19, 21–26	1,936.0	880.0	1,056.0
27–31	1,291.0	660.0	631.0
33–36	935.5	506.0	429.5
37–40, 32	972.5	638.0	334.5
41–44	1,075.0	404.8	670.2
TOTALS:	6,210.0	3,088.8	3,121.2

This percent loss was calculated based on guest level electrical work. The electrical work on the guest levels was relatively simple compared to the electrical work in the restaurants, kitchens, mechanical/electrical rooms, and public areas. So, it is reasonable that productivity losses in these areas were at least equal to the productivity losses on the guest levels. Thus, the electrical contractor's overall loss of productivity was at a minimum 61.5%.

Total Man-Hours Lost Due to Inefficiency

Compensation for T&M work was based on the actual number of hours spent on such work. Thus, the electrical contractor has already been compensated for any inefficiency in T&M work.

The total man-hours lost due to inefficiency were calculated as follows:

Direct Man-Hours Expended	-	412,311 MH
T&M Man-Hours Expended	-	119,309 MH
Man-Hours Subject to Inefficiency (Directed Man-Hours Expended – T&M Man-Hours)	=	293,002 MH
Man-Hours Lost Due to Inefficiency (293,002 $MH \times$ 61.5%)	=	180,196 MH

Thus, the electrical contractor lost 180,196 man-hours due to inefficiency during the period when problems occurred.

This case study reminds us, again, to develop as much detailed and accurate information as possible. It further illustrates that we must consider every possible factor that may come into play such as the learning curve. Perhaps most importantly, it alerts us to the requirement that we cannot just make a modification and expect that it works without verifying that the change we made (such as the vertical transportation) actually improved on the problem.

8

Industry Publications and Studies

When discussing inefficiency on a construction project, you will see reference made to various industry publications and studies. Normally, the reference to these studies is in the presentation of a claim or request for equitable adjustment because of increased costs related to loss of productivity. Far too often, the citations are treated as gospel with no analytical backup or verification. Even worse, these references are used to establish a quantitative basis for a claim, but the user has little to no understanding of the reference used, how it was developed, or its applicability to the situation.

Before anyone uses these documents as a basis for quantifying lost productivity, there should be a thorough understanding of the study being used, the development of the study and the reliability of that published information. It has been the authors experience that these studies are often used incorrectly in that they may not directly apply to the situation at hand, they have little, if any, quantifiable or analytical basis, or they do not clearly and accurately describe the actual application of the study. You may recall that we noted early on in this book that there are few absolutes in the area of construction productivity. The blind application of an industry published study inherently assumes the absolute applicability of the study to the existing problem. This may be far from reality. We have noted in this book that productivity in construction is a difficult area to define by formula or rules. Because tradesmen do not do the same tasks every day or even every hour, and because there are many factors that affect productivity, it is extremely difficult to do a statistically valid evaluation that will yield a simple formula applicable to most situations. There are many variables to consider, even if the comparison areas are being made on the same project. Naturally, the "studies" exacerbate those variables, as the amount of differing factors and considerations increase.

Ted Trauner, Chris Kay and Brian Furniss. Productivity in Construction Projects, (97–142) © 2022 Scrivener Publishing LLC

That does not mean that the studies performed are not worthwhile. They do serve to provide potential indicators of the effects of various factors on the productivity of a construction project. They have value as an indicator of where to look if problems are experienced. They have not yet offered valid formulas that can be blindly applied to every situation. This important distinction must be kept in mind when attempting to measure losses of productivity on a construction project.

It has also been the authors' experience that the use and expansion of industry studies will not go away. More will be created, more will be referenced, and more will be used. The reader is cautioned that just because a "study" was completed, or even accepted in the legal domain, that does not make it a valid study or analytical basis. The legal domain does not dictate engineering practice and principles. Would you trust a bridge more if it was deemed useful by the legal system, or a board of structural engineers and experts? Do you know whether the people making the legal decisions were provided ample information, analyses, or alternative options? Valid "studies" endure thorough peer review and the authors of these studies willingly provide the raw information behind the "study" available for consideration and scrutiny. It would be imprudent to trust that a car was deemed "safe" without performing research on the basis of the testing performed, the availability of the raw data from the tests, and then objectively reviewing and scrutinizing the results by the industry. We ask that the reader and analyst simply perform the same level of consideration before applying studies to their construction project, especially when considerable costs are at stake.

This chapter will look at the more common industry published standards, review the genesis of each, and comment on their applicability and use on the construction project. As each is discussed, the degree of weight that may be ascribed them should become obvious to the reader.

Bureau of Labor Statistics

One of the oldest publications that is often referred to is a study by the United States Department of Labor, Bureau of Labor Statistics (BLS). It is titled "Hours of Work and Output" and was issued in Bulletin No. 917 on May 21, 1947. The bulletin addresses the effects of overtime or an extended work week on the output of labor. Basically, this is a study of the effects of extended work periods on productivity.

The study was performed on manufacturing facilities and encompassed 78 different plants. It included 2,445 men and 1,060 women. The general results of the study indicate that as the hours of the work week increase over 40 hours, there is a reduction in the hourly productivity. The study further noted that an increased work week also resulted in increased absenteeism and increased incidence of injuries. In reviewing Bulletin No. 917, one may quantify various percentage levels of productivity or efficiency based on the case studies performed. The real question for us in the construction industry is "Can we rely on this study and use it to quantify perceived losses of efficiency on a construction project?"

Let's look closely at what this study presents.

First, this study was performed over 70 years ago. We believe it prudent to note that many things were different 70 years ago such as the individual work ethic, societal values, family priorities, equipment used, etc. Times have changed dramatically. Therefore, the sheer age of the study forces us to question its direct applicability to today's construction project.

Second, the study was conducted in a wartime (WWII) and immediate post-wartime environment. Reasonably, the wartime environment might well have had an effect on the motivation and performance of the workers, the stresses those workers were enduring, and other societal considerations that may not apply today.

Third, the study did not isolate variables in order to measure the affect that they might have on efficiency. For example, the study included operations where there was a wage incentive above 40 hours and also where there was no wage incentive above 40 hours. It noted that:

> *"It must be added, however, that long schedules had no adverse effect on workers operating under wage incentives if their pace during the shorter work schedule was moderate. Under such conditions they not only could maintain their efficiency at the longer hours but were even capable of improving it. It is likely, however, that this would not hold true for weekly hours above 60, because of the cumulative effects of fatigue."*

As can be seen from the preceding quote, some types of work with wage incentives above a 40-hour week, resulted in no loss of efficiency and a possible increase in efficiency. The study only *conjectured* that this might not hold true beyond a 60-hour week. It further noted that the effects of overtime were not consequential if the work was moderate as opposed to heavy. The study noted the differences among light, moderate, and heavy work as:

> *"In the discussion of each work schedule, findings will be shown separately for heavy, moderately heavy, and light operations. In the first category are operations such as are found in forge shops and foundries, and in which the work normally involves the handling of heavy materials. The category of light work involves, as a rule, the manual handling of materials up to about 5 pounds, or the mechanical handling of somewhat heavier objects. The moderately heavy group falls somewhere between these two."*

In the context of a construction project, how many operations require the manual handling of objects in excess of five pounds? Generally, not very many tasks require this. For example, an equipment operator is mechanically handling material and, therefore, would fall into the moderate category. Obviously, many tasks on a construction site would be categorized as moderate based on the definition given in the study. As a consequence, the more dire effects on productivity cited would not necessarily apply.

This study must be weighed in the context of exactly what was done and when it was performed, and does it provide a comparative basis to the modern construction project. The BLS study was only done in a manufacturing environment. Clearly, this is very different from a construction job site. It was performed over 70 years ago. Thus, we must recognize cultural and societal differences over time with respect to work. It applied primarily to piecework, which is vastly different than construction tasks, as piecework in a manufacturing context is far more monotonous and boring than performing varied tasks throughout the day. The average construction worker does not perform the same "piecework" during the day. The construction worker's time may vary among moving around the site, procuring material, getting tools, etc. Even seemingly monotonous construction tasks require a high degree of variability, such as height of work performed, varying levels of skill, and the project location where the work is performed. The manufacturing environment strives to eliminate variability in the operations and encourages repetitive work in order to maximize efficiency, whereas construction work can seldom be performed in the same manner.

Other findings of the study should also be noted. For example, it is often postulated that this study supports the position that work in excess of 40 hours in one week will cause a loss of efficiency. That is not an accurate conclusion. While some facilities that were studied supported that premise, others did not. For example, the study notes that:

> *"And it may be concluded that the addition of a sixth day without increasing daily hours had no adverse effect on efficiency."*

It further stated that:

> *"On the whole, it may be concluded that these studies indicate that the addition of a sixth day had no disadvantageous effect on output, provided daily hours were held to 8 per day."*

Therefore, it appears that while the overall study indicates situations in a manufacturing environment where overtime work may cause a loss of productivity, the application or method of applying the overtime hours directly relates to any effects on productivity.

When a careful review is made of the specific case studies used in the report, some very interesting results are observed. For example, in one of the case studies, which more closely approximated a construction environment, the results of a 58-hour week were surprising. The study notes:

> *"In study 9, 275 men were engaged in welding operations in the construction of various parts which were to go into battleships. The work involved considerable climbing about and permitted the use of individual initiative and resourcefulness. The work was not routinized, repetitive, or monotonous. Under these conditions, workers found it possible to maintain the same hourly efficiency under the 10-hour day and 58-hour week as they had under the 9-hour day and 45-hour week."*

From reading this, one is tempted to immediately conclude that for operations similar to some types of construction activities, a longer work week has no adverse effect on efficiency. It must be noted, however, that this case dealt with workers fabricating parts for battleships during a world war. It is very possible that pride and patriotism may have been a factor in this particular case. This is another example of the many variables that may affect the productivity on the project. While we cannot say definitely that pride and patriotism came into play, we cannot say it did not.

With respect to wage incentives, the study noted:

> *"...the men worked at straight day rates, and without any kind of wage incentive. In each case, output changed directly in proportion to the change in hours, indicating that the same pace was maintained regardless of hours. As output for men at piecework or other forms of direct incentives usually increased proportionately less than hours when they were raised substantially, the conclusion seems warranted that the slower pace which characterized the longer schedule for the men at straight day rates also prevailed during the shorter schedule. In short, operators who stood to gain nothing by working fast would not do so. Consequently, it was possible to increase hours – within reasonable limits – without impairing output. An hour's output resulted for each additional hour worked."*

The study continued, noting:

> *"Regardless of the reason for the speed of the work pace, it is obvious that individual incentive systems do not ipso facto mean that operators work at their optimum efficiency. Where the pace is moderate, therefore, output can be increased in direct proportion to the increase in hours, even though these are raised to 55, and perhaps to 60. This appears to be particularly true of operations which allow the operator considerable initiative and are not strictly routinized and monotonous. Although it was not possible to study such situations, instances were found in which the production of tool makers held up well under 60- and 70-hour weeks."*

Another area is worthy of note. Oftentimes it is suggested that once overtime operations are ceased or that there is a break for workers to become "refreshed," any adverse effects from overtime to that point will disappear and the work "will start anew." The BLS study noted that:

> *"...a shift to a new schedule did not result in an immediate change to readily ascertainable new patterns of efficiency, absenteeism, and output. It was found, for example, in case study 1, that it took about 4 months for the patterns of the 50-hour week to emerge clearly after a change from a 60-hour schedule. Similarly, in case study 7, the efficiency of the 40-hour week was maintained for 3 months after weekly hours had been raised to 50, and then decreased sharply to a lower level."*

A review of the underlying data in the BLS was analyzed using regression analysis. The goal was to use the information in the BLS to determine whether increased hours resulted in a decrease in efficiency. The BLS information available provided coefficients of determination (R^2) that were very weak when comparing the change in hours (independent variable) to the change in efficiency (dependent variable). The data in Tables 1 through 18 (in the BLS study) had an R^2 value of 0.09 when comparing the percentage change in hours to the % change in efficiency. In simple terms, this means that less than 10% of the efficiency change could be explained by the change in hours. Table 20 in the BLS, which discussed "relation of performance at various schedules of hours and incentive systems" had the highest R^2 value with a 0.59, meaning that 59% of the variation in efficiency could be explained by the changes in hours per week. Several other regression analyses were completed and the R^2 values remained around 10%. This substantiates that even if the BLS did somehow relate to modern construction, the data does not support that an increase in working hours caused a decrease in efficiency.

In summary, probably the only conclusion that one may draw from this study is that it indicates that productivity may be adversely affected if the work week is increased either in hours or days such that the total exceeds the 40-hour norm. It does not allow one to deduce or quantify the exact amount that productivity will decrease. To rely solely on this study to support any mathematical computation as a measure of lost efficiency because of overtime would not be a prudent course of action. Granted, this particular study does not offer formulas to quantify losses of productivity because of overtime. Others, however, do provide methods of calculation. These will be addressed next.

Business Roundtable Report

Another often cited "industry standard" is a bulletin by the Business Roundtable on the "Scheduled Overtime Effect on Construction Projects." The Business Roundtable is an organization of Chief Executive Officers of 200 large companies. The Business Roundtable has produced the results of studies it has had conducted on various topics. The overtime study is one of those reports. The study was published in 1974, updated in November 1980, and reprinted in 1986 and 1989.

In researching the history of this Business Roundtable Bulletin, it appears that the results published trace back to the early 1970s. There are two articles published by the American Association of Cost Engineers (AACE) in 1973. These articles address the same specific topics as shown in the Business Roundtable Bulletin. The genesis of these articles was from a study by the Construction Users Anti-Inflation Roundtable. Apparently, this was the precursor of the Business Roundtable. A review of the material presented in these articles shows that the graphic and tabular information presented is identical to that contained in the Business Roundtable Bulletin. Therefore, the study had to have been performed prior to 1974, which makes it at least 45 years old. As was noted in our discussion of the data from the Bureau of Labor Statistics, many changes have occurred in the construction industry and in the work environment in this country. Thus, as noted earlier, the age of the study must give rise to its direct applicability in today's construction environment.

The study as summarized in the AACE bulletins was based on a review of construction work at Proctor & Gamble. No explanation was located as to the exact nature of the projects, the exact number of projects, or the exact craftsmen involved. It is reasonable to presume that since the work was for a manufacturing operation, the type of construction would be to some degree vertical and predominately it would be process work. William Schwartzkopf in his book *Calculating Lost Labor Productivity in Construction Claims* notes that:

> *"The Business Roundtable report is based entirely on data from a single project constructed in Green Bay, Wisconsin, for Proctor & Gamble. The data is from a series of short jobs over a ten-year period. The entire study period was a period with excellent labor-management relations. The data was calculated using fixed-unit rate standards. As a result, the data did not yield a direct comparison between actual overtime and straight time productivity."*

It is particularly significant that the AACE article clearly indicated that:

> *"Sources of quantitative data include Bulletin No. 917, Department of Labor, Bureau of Labor Statistics (1947) and two relatively recent evaluations by Mechanical Contractors Association of America, Bulletin No. 18A, and the National Electrical Contractors Association's Southeastern Michigan Chapter, and experience records of Proctor & Gamble construction operations."*

It is unknown how the information was integrated into the various tables and graphs presented in the Business Roundtable Bulletin. It must be noted, however, that as was seen from the preceding discussion of the Bureau of Labor Statistics and will be seen in the later discussion of the Mechanical Contractors Association and the National Electrical Contractors Association, the information used as a basis may be neither accurate nor applicable to the vast majority of construction projects. The supporting data does not justify the use of the Business Roundtable Bulletin as a quantitative measure of loss of efficiency due to the effects of overtime.

In reading the Business Roundtable Bulletin, the conclusions reached are quite clear that overtime can adversely affect the productivity on a construction project. The results are presented both graphically and in tabular format. Figure 8.1 shows the Roundtable curves for the cumulative effects of overtime over an extended period for 50- and 60-hour work weeks.

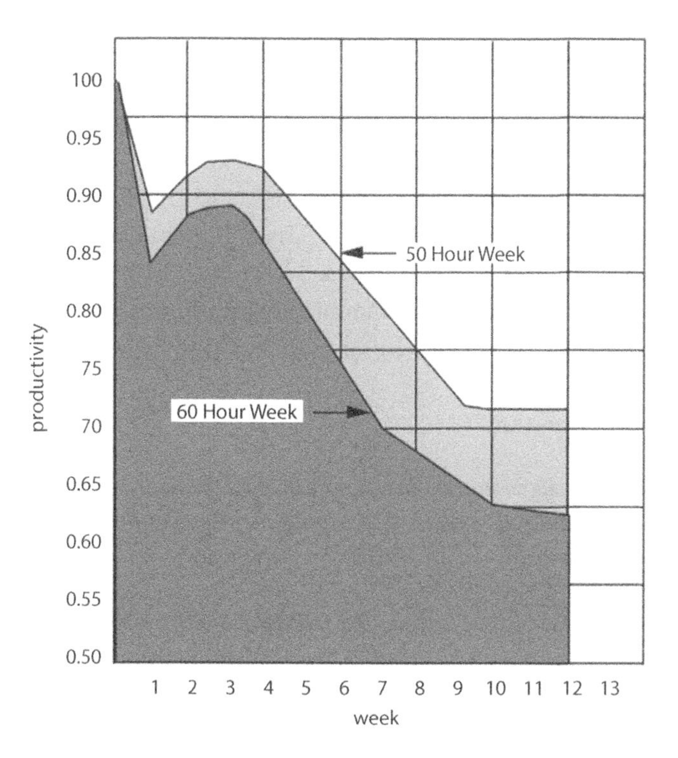

Figure 8.1 Cumulative effect of overtime on productivity for a 50- and 60-hour workweek; BRT, page 10.

1	2	3	4	5	6	7	8
	Productivity	Rate					
50 Hour Overtime Work Weeks	40 Hr. Week	50 Hr. Week	Actual Hour Output for 50 hr. Week	Hour Gain Over 40 hr. Week	Hour Loss Due to Productivity Drop	Premium Hours	Hour Cost of Overtime Operation (at 2X)
0-1-2	1.00	.926	46.3	6.3	3.7	10.0	13.7
2-3-4		.90	45.0	5.0	5.0	10.0	15.0
4-5-6		.87	43.5	3.5	6.5	10.0	16.5
6-7-8		.80	40.0	0.0	10.0	10.0	20.0
8-9-10		.752	37.6	-2.4	12.4	10.0	22.4
>10		.750	37.5	-2.5	12.5	10.0	22.5

Figure 8.2 Relationships of hours worked, productivity and costs (40 hours vs. 50 hours); BRT, page 12.

Figure 8.2 shows a tabular representation of the relationship of productivity and cost for a 40- and 50-hour workweek.

In both figures, one can clearly see that the study concludes that productivity is adversely affected as a result of the hours in excess of 40 hours per week and also affected by the duration over which the overtime extends.

Once again, let's look at the basis of the report.

The references given at the end of the report do not include any studies dealing with construction projects. Note that two of the references are the Mechanical Contractors' Association of America Bulletin 18-A, "How Much Does Overtime Really Cost?" and the National Electrical Contractors' Association publication "The Hidden Cost of Overtime." Both of these "studies" will be addressed later in this chapter. The remaining references are primarily for manufacturing situations. Note also, that one reference is the Bureau of Labor Statistics study which was discussed previously.

The report notes that:

> *"This paper relates only to operations where the total job is placed on an overtime basis for an extended period of time. Meaningful data to cover periodic overtime is not available."*

This statement seems curious. If a job has overtime only for one trade but not all trades, are we to assume that the report does not apply? One would conclude that from the preceding quote. Therefore, the use of this report for validation of lost productivity becomes questionable.

The Business Roundtable Report notes that:

"...the following conclusions have been reached:

For hours above eight per day and 48 per week, it usually took three hours of work to produce two additional hours of output when the work was light. For heavy work, it took two hours to produce one hour of additional output."

Yet this conclusion is footnoted to the BLS study that was previously discussed. Those conclusions may apply but only in the context of the BLS study. To represent that as a conclusion of the BLS study misrepresents to some extent what is contained in the entire document, as other parts of the BLS also contradicted that conclusion.

When one reviews the Business Roundtable Bulletin in detail, there are several subtleties and inconsistencies which become apparent. A few of these will be noted as follows to demonstrate the risk one accepts when attempting to use this bulletin in any quantitative fashion.

Figure 8.1, as shown previously, was replicated from the Business Roundtable Bulletin. In that bulletin it is identified as Figure 8.2. The bulletin then shows its Figure 8.4, which was shown previously as Figure 8.2. The representation of this table shown in Figure 8.2 is that it is a quantification of the graphs in Figure 8.1. Unfortunately, the numbers in the table do not directly correspond with what is shown in the graphs. The Business Roundtable Bulletin very subtly notes this anomaly when it states:

"Column 3 reflects an interpretation of the productivity rate from Figure 2 for the periods shown in Column 1."

If the graphs are scaled based on the units shown, you will not arrive at the same numbers as shown in the table. Therefore, this "interpretation" raises serious questions as to the propriety of using the numbers presented in the table for any meaningful calculation of lost efficiency.

In one sentence on page 13 of the Business Roundtable Bulletin it states that:

"This results from the reduced productivity applying to a smaller base of overtime hours and indicates that a 45-hour job schedule very quickly becomes nothing more than wage inflation."

This sentence is related to a discussion of how productivity was reduced to a "point of no return" in 50- and 60-hour workweeks. The AACE articles were more explicit in this discussion. It noted that earlier studies indicated:

> *"...the workman to be 12% more productive for 4 weeks at an 8-hour day than he is for 4 weeks at a 9-hour day."*

In other words, the bulletin very quietly indicates that for overtime less than a 50-hour week results in a loss of efficiency even greater than indicated in the studies for the 50-hour and 60-hour workweek.

A July 2006 article in *Cost Engineering* by Seals and Rodriquez was critical of the Business Roundtable study. The article asserted that the BRT Report was "fundamentally flawed" and "contained significant methodological and theoretical errors." The article further noted that the BRT Report contained three significant failures:

1. It contained no actual data or explanation of how data was collected, filtered, or manipulated.
2. There was a lack of correlation between the curves and the source material.
3. Data was not collected following set standards, the study was too specific to be applied universally, did not make an explicit connection between fatigue and efficiency, and was outdated.

This article is worth reading and is in line with the criticisms expressed in this book.

Because the Business Roundtable Bulletin is based on non-scientific studies such as the MCAA and NECA, because it draws in part from the Bureau of Labor Statistics manufacturing study, and because it is limited to a study only of Proctor & Gamble for any independent assessment, its reliability is highly questionable. It appears that this bulletin has no value as a quantitative measure of loss of productivity for construction operations for an overtime situation. Its sole use is additional support that overtime may cause a reduction in the productive output of construction workers. The amount of that reduction should not be quantified from the information presented in this bulletin.

National Electrical Contractors Association

The National Electrical Contractors Association has published numerous documents related to the subject of productivity in construction. Apparently, NECA has two alter organizations, the Electrical Contracting Foundation and ELECTRI'21. The following is a partial list of the NECA publications:

- Quantifying the Cumulative Impact of Change Orders
- The Effect of Temperature on Productivity
- Factors Affecting Labor Productivity
- Project Peak Workforce
- Manpower Loading: The Rate of Manpower Consumption
- Guide to Electrical Contractors' Claims Management
- Overtime and Productivity in Electrical Construction
- What-To-Do-Guide for Schedule Acceleration and Compression
- Productivity Enhancement Focusing on Labor Efficiency
- Normal Project Duration
- Rate of Manpower Consumption
- The Effect of Multistory Buildings on Productivity
- Impact of Change Orders on Labor Efficiency
- Stacking of Trades
- Negotiating Loss of Labor Efficiency
- Strategies for Minimizing the Economic Consequences of Schedule Acceleration and Compression

Our discussion will be limited to just a couple of the publications, though it should be noted that many of the assessments made concerning these publications apply to most of the other publications.

The National Electrical Contractors Association (NECA) published a document in 1969 titled "Overtime and Productivity in Electrical Construction." The second edition of this was reprinted in 1989. While first published in 1969, the document notes that the "study" was conducted ten years earlier or approximately in 1959. This means that the information used as the basis for this publication is now over 60 years old. Once again, many things have changed in the past 60 years. This obviously brings into question the applicability of the data today; 60+ years later.

Schwartzkopf notes that:

> *"This study is based on a survey of NECA membership and, as a result, is not based upon empirical or field study."*

This statement is not exactly correct. The NECA publication notes that it is based on a survey, as Schwartzkopf points out:

> *"Some years ago, NECA published the results of a survey of its membership regarding the experience of electrical contractors with reduced productivity associated with overtime. The survey consisted of four questions concerning overtime on a sporadic, short-duration basis, and two questions concerning continuous application of overtime over several successive weeks."*

The NECA publication, however, goes on to note that:

> *"A study conducted by the NECA Southeastern Michigan Chapter of jobs worked during 1964 bears this observation out dramatically. The findings of the study are presented..."*

Therefore, the data contained in the NECA publication includes its survey results and the results of the chapter study. The results of that study are shown in Figure 8.3. It should be noted that this study reflects a period up to 28 days or four weeks. Additional charts contained in the NECA publication extend out to 16 weeks and apparently are the results of the survey information rather than the Chapter study. Figure 8.4 and Figure 8.5 show the results of the survey information for a 50-hour workweek. As can be seen, this approach yields a range of possible losses in productivity from 0% to 37%. The figure also does not illustrate a "point of no return" as was hypothesized by the Business Roundtable Bulletin.

If one reads the entire NECA publication, he/she will note that NECA concludes that the information it has acquired correlates quite closely with the data presented by the Bureau of Labor Statistics study from the mid-1940s. The information from both the Bureau of Labor Statistics and the NECA research is summarized in various charts and graphs throughout the publication. Perhaps the most significant point made in the NECA publication is the following:

Days and Hours	Weekly Hours	Percent Productivity			
		7 Days	14 Days	21 Days	28 Days
Six 9's	54	96–94%	94–91%	92–86%	90–85%
Six 10's	60	93–91%	89–86%	86–82%	82–77%
Six 12's	72	88–86%	82–79%	76–72%	70–65%
Seven 8's	56	91–89%	86–84%	82–76%	77–73%
Seven 9's	63	89–87%	83–80%	78–74%	72–67%
Seven 10's	70	86–84%	79–76%	72–68%	65–60%
Seven 12's	84	80–78%	70–67%	60–56%	50–45%

Figure 8.3 Tabular results of the Southeastern Michigan study.

	Productivity			Productivity Loss		
	Low	Average	High	Low	Average	High
Week #1	95%	98%	100%	0%	2%	5%
Week #2	92%	95%	97%	3%	5%	8%
Week #3	89%	82%	94%	6%	8%	11%
Week #4	86%	89%	91%	9%	11%	14%
Week #5	83%	86%	88%	12%	14%	17%
Week #6	80%	83%	85%	15%	17%	20%
Week #7	77%	80%	82%	18%	20%	23%
Week #8	74%	77%	79%	21%	23%	26%
Week #9	71%	74%	76%	24%	26%	29%
Week #10	69%	72%	74%	26%	28%	31%
Week #11	68%	71%	73%	27%	29%	32%
Week #12	67%	70%	72%	28%	30%	33%
Week #13	66%	69%	71%	29%	31%	34%
Week #14	65%	68%	70%	30%	32%	35%
Week #15	64%	67%	69%	31%	33%	36%
Week #16	63%	66%	68%	32%	34%	37%

Table 1
5 Days & 10 Hour Work Days

Figure 8.4 Southeastern Michigan Study survey information for a 50-hour workweek.

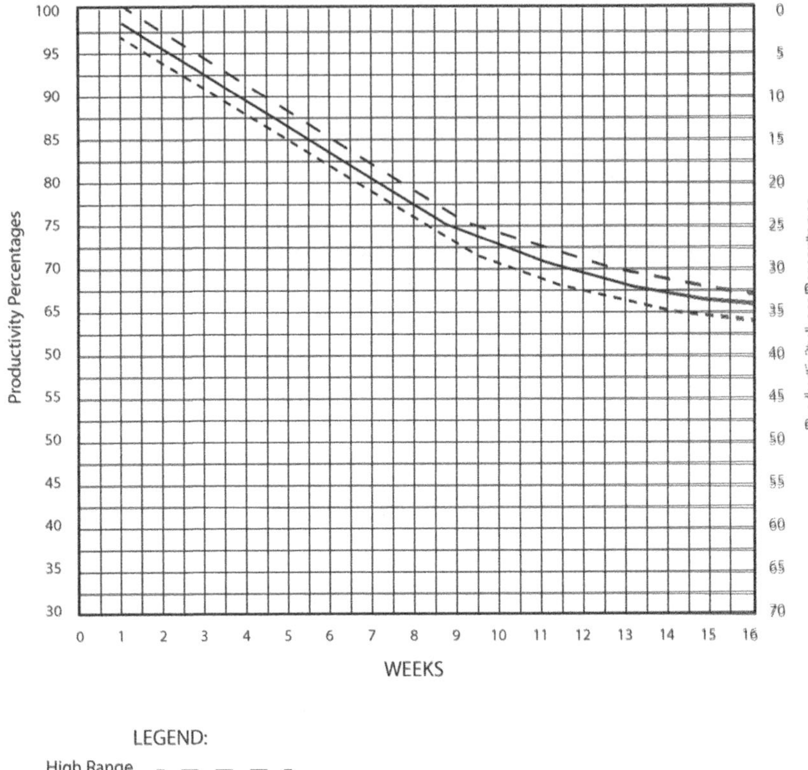

Figure 8.5 Southeastern Michigan study survey plot.

"The need for thorough, orderly documentation of project expenses cannot be overstressed. The most persuasive argument which can be raised....is to show that the proposed loading factors are based upon the experience of one's own organization, and that they are not out of line with the experience of the industry as a whole.

Supporting a claim requires in addition that the particular project in question be thoroughly documented. Courts and boards are rarely willing to apply purported industry averages to the adjudication of a claim directly, although they may reasonably be expected to use such averages as benchmarks against which to judge the reasonableness of a specific claim.

Therefore, it is important to keep both good current records and orderly historical files."

In simple terms, NECA points out that its members should not be using the "industry standards" as the primary method to support a claim but rather to base a claim on the detailed project records and the company's detailed historical records. This concept will be addressed in other sections of this book.

More recently, NECA produced new studies related to productivity in electrical construction. The publications have been noted previously.

Of the preceding publications, the Factors Affecting Labor Productivity for Electrical Contractors incorporates the results of some of the other studies. For that reason, it will be addressed first.

Review of Factors Affecting Labor Productivity for Electrical Contractors

This study attempts to quantify loss of efficiency for seven items:

- Overtime
- Overmanning
- Shift Work
- Stacking of Trades
- Owner-Furnished Items
- Beneficial Occupancy
- Cumulative Impact or Ripple Effect

The study was based on data provided by contractors and collected between 1999 and 2002 for 152 projects, including commercial, institutional, and industrial projects. The projects were competitively bid on in a design-bid-build delivery system. While the report notes the number of projects, it does not indicate that every project was utilized for each area where formulas were developed. For example, it is unclear if all 152 projects had usable data for the derivation of the formula for overtime loss of productivity. The same is true for the other six items in the study. Therefore, the report fails to define the exact data base for each specific item.

For all seven items a formula is developed based on a regression analysis of the available information. The authors, however, fail to provide the raw data used for any analysis, and fail, except in one case, to provide any measure of the reliability of the equations developed. Only one value for R^2 is provided and that relates to the overtime study. In that instance, the R^2 value is 53.4%. This means that the reliability of this calculation is about 3.4% more reliable than the flip of a coin (50% of the time you get "heads"). Though the specific R^2 values were requested from the NECA and the author of the study, this information has not been provided. If this one value is indicative of the entire study, then the results have questionable reliability.

The new publications contain various formulas for quantifying the amount of lost productivity for various causes. Unfortunately, the details of the research are not clearly described in the publications. The authors of this book have written to the NECA and ELECTRI'21 seeking additional information to allow a fair peer review of the published material. At the time of publishing this book, no response was received. Therefore, the discussion in this book is based on the material as presented in the NECA publications. It should be noted that the lack of detailed information for the formulas developed precludes any meaningful and analytical review to assess the validity of the methods used, the reliability of the usable information, and the merit of any conclusions reached. Until such time as the NECA and ELECTRI'21 provide backup information for these publications, they should not be considered reliable or accepted as a reasonable estimate for any alleged losses in productivity.

Mechanical Contractors Association of America

The Mechanical Contractors Association (MCAA) offers a publication titled *Change Orders, Productivity, Overtime—A Primer for the Construction Industry*. It was first published in 1974, and there have been multiple revisions since then that generally use the same information. This publication offers comments on the effects of change orders and overtime on the productivity of the contractor. In its section on "Productivity," the MCAA publication includes tables providing various factors affecting labor productivity. This table of factors is often used in support of claims for lost efficiency.

In the discussion of overtime, the MCAA refers to various studies such as the Construction Industry Cost Effectiveness (CICE) Project, the Business Roundtable, etc. Various charts and graphs are included in the MCAA publication that were taken from the Bureau of Labor Statistics and the Corps of Engineers Modification Impact Evaluation Guide. The MCAA generated the tables shown in Figure 8.6 that represent loss of efficiency due to overtime ranging from 3 1/3% to 29%.

Section PD2 of the publication summarizes the MCAA guidelines for losses in productivity in a 2-page table which provides a range of percent loss for various factors. This table is shown in Figure 8.6.

As can be seen in Figure 8.6, the factors affecting productivity include stacking of trades, morale and attitude, reassignment of manpower, crew size inefficiency, concurrent operations, dilution of supervision, learning curve, errors and omissions, beneficial occupancy, joint occupancy, site access, logistics, fatigue, ripple, overtime, and season and weather change. The represented potential losses in productivity range from 1% to as high as 50%.

Oftentimes, the factors shown in Figure 8.6 are used in the presentation of a claim for loss of productivity. Unfortunately, there is no clear instruction as to how one correctly applies these factors. It is immediately apparent that the assignment of these factors is totally subjective. Therefore, one can reach any desired percent loss of productivity merely by the method of assigning the factors in the manner he/she chooses or deems appropriate.

	Percent of Loss if Condition:		
Factor	Minor	Average	Severe
1. **STACKING OF TRADES**: Operations take place within physically limited space with other contractors. Results in congestion of personnel, inability to locate tools conveniently, increased loss of tools, additional safety hazzard and increased visitors. Optimum crew size cannot be utilized.	10%	20%	30%
2. **MORALE AND ATTITUDE**: Excessive hazzard, competition for overtime, over-inspection, multiple contract changes and rework, disruption of labor rhythm and scheduling, poor site conditions, etc.	5%	15%	30%

Figure 8.6 Factors affecting labor productivity; MCAA 1994, Productivity PD2.

(Continued)

Factor	Percent of Loss if Condition:		
	Minor	Average	Severe
3. **REASSIGNMENT OF MANPOWER**: Loss occurs with move-on, move-off men because of unexpected changes, excessive changes, or demand made of expedite or reschedule completion of certain work phases. Preparation not possible for orderly change.	5%	10%	15%
4. **CREW SIZE INEFFICIENCY**: Additional men to existing crews "breaks up" original team effort, affect labor rhythm. Applies to basic contract hours also.	10%	20%	30%
5. **CONCURRENT OPERATIONS**: Stacking of this contractor's own force. Effect of adding operation to already planned sequence of operations. Unless gradual and controlled implementation of additional operations made, factor will apply to all remaining and proposed contract hours.	5%	15%	25%
6. **DILUTION OF SUPERVISION**: Applies to both basic contract and proposed change. Supervision must be diverted to (a) analyze and plan change, (b) stop and replan affected work, (c) take off, order and expedite material and equipment, (d) incorporate change into schedule, (e) instruct foreman and journeyman, (f) supervise work in progress, and (g) revise punch lists, testing and start-up requirements.	10%	15%	25%
7. **LEARNING CURVE**: Period of orientation in order to become familiar with changed condition. If new men are added to project, effects more severe as they learn tool locations, work procedures, etc. Turnover of crew.	5%	15%	30%
8. **ERRORS AND OMISSIONS**: Increases in errors and omissions because changes usually performed on crash basis out of sequence or cause dilution of supervision or any other negative factors.	1%	3%	6%

Figure 8.6 (Continued) Factors affecting labor productivity; MCAA 1994, Productivity PD2.

(Continued)

Factor	Percent of Loss if Condition:		
	Minor	Average	Severe
9. **BENEFICIAL OCCUPANCY**: Working over, around or in close proximity to owner's personnel or production equipment. Also badging, noise limitations, dust and special safety requirements and access restrictions because of owner. Using premises by owners prior to contract complerion.	15%	25%	40%
10. **JOINT OCCUPANCY**: Change causes work to be performed while facility occupied by other trades and not anticipated under original bid.	5%	12%	20%
11. **SITE ACCESS**: Interferences with convenient access to work areas, poor man-lift management or large congested worksites.	5%	12%	30%
12. **LOGISTICS**: Owner-furnished materials and problems of dealing with his storehouse people, no control over material flow to work areas. Also contract changes causing problems of procurement and delivery of materials and rehandling of substituted materials at site.	10%	25%	50%
13. **FATIGUE**: Unsual physical exertion. If on change order work and men return to base contract work, effects also affect performance on base contract.	10%	15%	20%
14. **RIPPLE**: Changes in other trades' work affecting our work such as alteration of our schedule. A solution is to request, at first job meeting, that all change notices/bulletins be sent to our Contract Manager.	10%	15%	20%
15. **OVERTIME**: Lowers work output and effiency through physical fatigue and poor mental attitude.	10%	15%	20%
16. **SEASON AND WEATHER CHANGE**: Either very hot or very cold weather.	10%	20%	30%

Figure 8.6 (Continued) Factors affecting labor productivity; MCAA 1994, Productivity PD2.

The most significant question about this table in Figure 8.6 is its origin. A few years ago, the authors contacted the MCAA in an attempt to determine the basis for the factors. The e-mail response noted that the information was so old that the MCAA was unaware of the exact source. In a subsequent investigation of the MCAA factors during a trial before the Department of Veterans Affairs Board of Contract Appeals, the MCAA executive vice president submitted a sworn written statement setting forth the history of the MCAA factors. Two significant statements from that sworn statement are worthy of note. The first notes:

> *"Because MCAA and its membership recognize that the loss of labor productivity is difficult to quantify with specificity, the MCAA Factors are expressly intended to serve only as a point of reference for mechanical contractors and other parties. The specific percentage values set forth in the MCAA Factors must be applied with careful consideration and a review of the facts surrounding the loss of productivity. The MCAA Factors are intended to be used in conjunction with the experience of the particular contractor seeking to use them, because the percentage of increased costs could well vary from contractor to contractor, crew to crew, and job to job."*

The sworn statement most significantly states that the factors were developed beginning in the late 1960s and continued into the early 1970s. With respect to the basis of the factors, the sworn statement notes:

> *"To the best of the MCAA's current knowledge, the information contained in the MCAA Factors was gathered anecdotally from a number of highly experienced members of the MCAA's Management Methods Committee. MCAA does not have in its possession any records indicating that a statistical or other type of empirical study was undertaken in order to determine the specific factors, or the percentages of loss associated with the individual factors."*

Therefore, the use of the MCAA Factors in the support of a claim seems questionable at best. There appears to be no scientific or empirical basis for the factors but rather the anecdotal assessment or estimate of members of a committee.

The Leonard Study

The Leonard Study is often referenced to support a loss of inefficiency caused by change orders on a project. Not only is it used to support that inefficiency occurred but also to provide percentages of loss to the project. Before one accepts this study as reliable, the basis for the work should be understood. The following discussion will provide useful background on the study and allow the reader to decide the applicability of the study to an analysis of inefficiency.

The Leonard Study was published in August 1988. It was a thesis for a master's degree at Concordia University, Montreal, Quebec, Canada.

Perhaps the most efficient way to start a discussion of this study is to summarize the conclusions presented in the study. In Chapter VI of the study, Leonard draws 13 conclusions that can be summarized as follows:

1. Most change orders are caused by design errors and omissions.
2. Individual change orders that disrupt and delay work adversely affect productivity.
3. Change orders have a ripple effect on the productivity of unchanged work.
4. When change order labor hours exceed 10% to 15% of the original contract work, there is a detrimental effect on productivity.
5. Causes of productivity losses caused by change orders include: stop-and-go operations, out-of-sequence work, loss of productive rhythm, demotivation of workers, learning curve losses, unbalanced crews, excessive manpower fluctuations, unbalancing of successive operations, lack of management support, and acceleration if time extensions are not granted.
6. Change order productivity losses are primarily experienced later in the job. As a consequence, delays may occur.
7. Project completion dates may be delayed significantly as a result of disruptions, additional work, and losses in productivity.

8. Scheduling and coordination of the work are significant factors in productivity. Change orders may make this more difficult but also more important.
9. Individual change orders may affect productivity depending on when work can proceed relative to the planned start of an activity.
10. Productivity losses are best calculated after the fact by the use of a measured mile analysis ("differential method of cost calculation").
11. The relationship between change orders and productivity loss is linear and the correlations are "relatively" strong.
12. The amount of productivity loss from change orders is affected by the type of work but not the type of construction.
13. Other causes of productivity losses (separate from change orders) can increase the amount of lost productivity.

Before accepting Leonard's conclusions, it is appropriate to understand the basis for the study and the methodology used in the study. The most important questions that must be addressed are:

- What is the basis of the study?
- Does the study analytically support the conclusions reached?

Let's begin with the basis of the Leonard study.

Leonard begins the study by noting that 90 cases from 57 projects were studied. One must question if this is truly 90 cases or more appropriately 57 cases, since multiple "cases" on the same project should show the same results. If so, this can skew or bias the results of the study by providing more weight to results from a project with multiple cases as compared to a project with a single case.

Most importantly, Leonard also notes that this study is not meant to replace a measured mile and, in fact, the measured mile is the preferred method for assessing lost productivity. Leonard clearly states that this study is a method for *estimating* lost productivity when a measured mile is not possible.

Leonard contends that this study is equivalent to an industry-wide study because it is based on actual projects. This assertion is highly questionable and will be addressed a bit more later on.

Leonard notes that his study used data from 84 separate contracts on 57 independent projects comprised of buildings and industrial facilities. No heavy/highway work was included.

Perhaps the most important aspect of the study setup was how the data was acquired. Leonard gathered information from a claims consulting firm for all the projects. Furthermore, Leonard states that:

> *"...84 contracts were identified on which the contractor had experienced loss of productivity as a result of change orders."*

Recognize that this means that someone else had already concluded that change orders caused loss of productivity. If you accept that this identification was correct, then the study is merely an attempt to quantify that loss. There is no indication that the losses in productivity may have had other factors that influenced the magnitude of the loss. Because of this huge assumption, the validity of the study must be questioned. Furthermore, to objectively determine whether change orders cause productively losses, an analyst would need to not only analyze projects with alleged inefficiency losses due to change orders, but it would also need to analyze projects without inefficiency losses where change orders occurred. To assume that change orders lead to productivity losses would be biased and skew any alleged quantifiable link between change orders and productivity.

The Leonard Study information that was utilized was gathered from the following sources:

- Contractors' Claims
- Claim Evaluations
- Expert Reports
- Job Files

There was no independent analysis by Leonard to determine if change orders caused any loss of productivity. Rather, it was accepted that because the firm supplying the information believed change orders caused a loss, then it was a fact. A highly questionable premise for an analysis.

Leonard also states that in many cases information was not available on the actual man-hours spent on change orders. Instead, estimates by the contractor or the owner were utilized to define man-hours for the study. This is yet another highly questionable premise. Basically, there is no accurate record of the man-hours expended on change orders, yet it is the basis for the statistical gymnastics that followed.

Leonard also does not independently measure the losses in productivity. Instead, the study accepts whatever measure was assigned by an expert report outside of the Leonard Study. In cases where there was no expert report, Leonard used either some form of measured mile or total cost/modified total cost. As a consequence of this, one must conclude that this study has no accurate or reliable measure of the productivity losses that are used in subsequent statistical calculations. Leonard further points out that in 15% of the cases; progress was over-reported early in the job. This resulted in the remaining work being understated and the early productivity achieved as being overstated. Hence, the data was inaccurate to begin with. Yet Leonard does not state whether these cases were excluded from the study.

After gathering data, Leonard then performs a series of statistical calculations based on regression analysis. Some of these show a reasonable level of correlation, while others don't. Interestingly, Leonard does not have a large amount of data points in any of the scatter diagrams in the study. Based on a visual review of the figures in the study, the data points range from a high of 28 to a low of eight. Using eight data points to draw conclusions for future estimates is highly questionable.

The results of the regression analysis yield correlation coefficients (R) that are at best suggestive of a correlation. Leonard fails to take the one further step and calculate the coefficient of determination (R^2). This calculation would be a better measure of the reliability of the analysis for making predictions. A quick calculation of the coefficient of determination for all of Leonard's regression calculations shows that the reliability of this study for predicting loss of productivity is less than the study implies. For example, Leonard's Civil/Architectural contracts, Type 2, have a correlation coefficient of 0.74. This yields a coefficient of determination of 0.55. That means that the model Leonard developed may explain only 55% of the variation of productivity versus the man-hours on change orders. The other 45% of variation is unexplained and resulted from a factor other than change orders. It is highly questionable to use a model that can at best account for 55% of the variation as an accurate method for estimating loss of productivity. Yet Leonard states that the 0.74 coefficient of correlation indicates a "relatively strong correlation." Simply put, this is incorrect.

Finally, it must be noted that the overall conclusions in the study are not supported by the analytical work that is performed. Though some relationships appear to exist in the statistical models generated, they are based on questionable data. Regardless of the results of the statistics, the conclusions are far broader than the analysis supports.

Corps of Engineers Modification Impact Evaluation Guide

The U.S. Army Corps of Engineers published its Modification Impact Evaluation Guide, EP 415-1-3, in July 1979. The purpose of the Guide is stated as:

> *"This pamphlet provides information and guidance on the identification and evaluation of that portion of the fixed-price construction contract modification defined as impact on the unchanged work."*

The Guide material recognizes the possibility that certain occurrences on a construction project may give rise to a reduction in the productivity achieved on the project. Both equipment and labor losses of productivity are addressed in the Guide. With respect to equipment, the Guide notes that:

> *"Reduction in productivity because of equipment crowding or increased traveling time requires a study of the individual situation. The objective of such a study is to define the production time lost to traveling (hours) and the loss of productivity caused by crowding (converted to hours), plus equivalent additional costs for operators and oilers (when applicable)."*

With respect to manpower, the Guide addresses this in much more detail. Several factors are identified which may adversely affect the productivity of the labor force on the project. These factors include disruption, crowding, acceleration, increased crew sizes, increased hours worked, multiple shifts, and morale. For several of these factors, the Guide presents graphs as a guide for estimating lost productivity. It should be kept in mind that the Guide was established to assist Corps of Engineers personnel in the upfront resolution of changes. As a consequence, the graphs and charts included are not presented as scientific studies but rather as "information or trends."

The narrative for each of the factors noted is instructive since the Corps, as one of the largest construction owners in the world, is recognizing how these factors can adversely affect productivity. In the discussion of disruption, the Corps notes:

> *"Disruption occurs when workers are prematurely moved from one assigned task to another. Regardless of the competency of the workers involved, some loss in productivity is inevitable during a period of orientation to a new assignment. This loss is repeated if workers are later returned to their original job assignment."*

The Guide then offers various learning curves based on industrial tasks to allow some estimate of the effect of disruptions. The following Figures 8.7 and 8.8 show the learning curve data presented in the Guide.

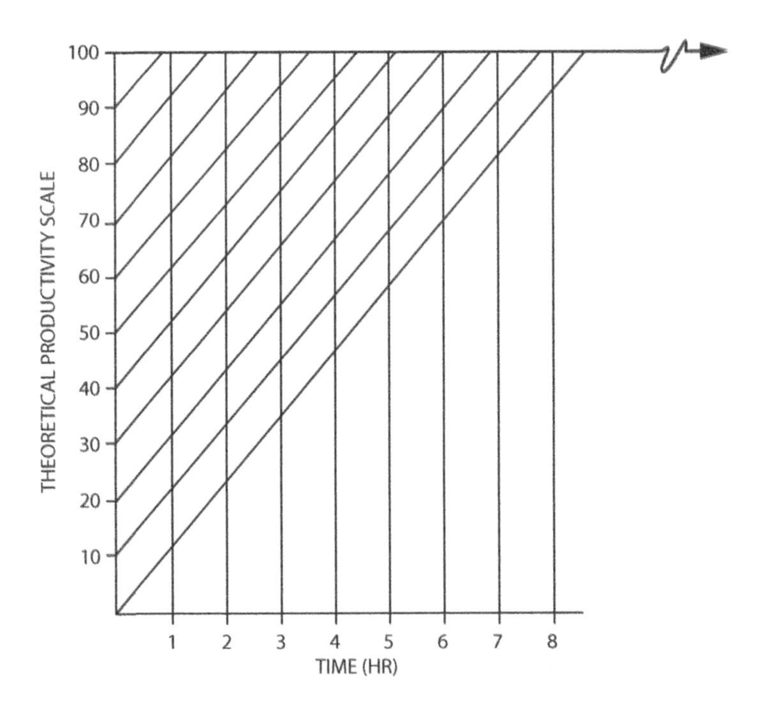

Figure 8.7 EP 415-1-3, July 2, 1979; Construction operations orientation/learning chart.

(BASED ON CONSTRUTION OPERATIONS ORIENTATION/LEARNING CHART)

PRODUCTIVITY STARTING POINT	DURATION (HR)	AVERAGE LOSS (HR)
100	0	0
90	0.8	0.4
80	1.6	0.8
70	2.4	1.2
60	3.2	1.6
50	4.0	2.0
40	4.8	2.4
30	5.6	2.8
20	6.4	3.2
10	7.2	3.6
0	8.0	4.0

Figure 8.8 EP 415-1-3, July 2, 1979; Productivity losses derived from Figure 8.7.

The Guide also recognizes that crowding can affect productivity. It notes, however, that merely because more activities are worked at once does not necessarily mean that crowding will be a factor. It clarifies this by noting that *"both increased activities stacking and limited (congested) working space must be present for crowding to become an impact cost."* Therefore, it is important to assess the actual project conditions before assuming that crowding has occurred and has had any effect on productivity.

In the area of acceleration, the Guide lists factors as subsets of this. These include increased crew sizes, increased hours worked, and multiple shifts.

For increased crew sizes the Guide notes that the optimum crew size is one that has the minimum workers to perform the task in the allocated time frame. As workers are added, a lesser return is seen on the production achieved. In other words, production increases are not linear as we increase the size of a crew. The Guide provides charts to estimate the effects of increased crew sizes. The major problem which is not clearly addressed in the Guide is the determination of the optimal crew size such that the charts can be used. It is also unclear as to how these specific charts were developed.

For increased work hours, shifts over 8 hours per day and more than 5 days per week, the Guide offers the composite chart in Figure 8.8 showing losses in productivity ranging from 2% to 40%:

Figure 8.8 was based on information for a four-week period of time. The Guide notes that it is assumed that the loss of productivity beyond four weeks will flatten out, implying no further losses after a four-week period.

The Guide notes multiple shifts as an increase in cost but does not address how one approaches the quantification of these costs. It does point out that lighting, cold weather, etc., are factors which come into play when multiple shifts are worked.

Finally, when the Guide addresses the area of morale, it notes that workforce motivation is the responsibility of the contractor. While it recognizes that morale has an influence on productivity, it postulates that "*The degree to which this may affect productivity, and consequently the cost of performing the work, would normally be very minor when compared to the other causes of productivity losses.*"

In a letter dated June 14, 1996, the Corps of Engineers "hereby rescinded" the use of EP 415-1-3. To the authors' knowledge, the Corps of Engineers has not reinstated the use of EP 415-1-3.

Construction Industry Institute

The Effects of Scheduled Overtime and Shift Schedule on Construction Craft Productivity

In December 1988, the Construction Industry Institute published a study of overtime and shift work on craftsmen productivity. The results of that study vary markedly from the previous studies noted. The study reaches four main conclusions, which are:

1. Previous studies by BLS, the Business Roundtable, and others are not consistent predictors of productivity loss during overtime schedules for construction projects in this study.
2. Even on the same project working an overtime schedule, productivity trends of individual crews are not consistent.

3. Productivity does not necessarily decrease with an overtime schedule.
4. Absenteeism and accidents do not necessarily increase under overtime conditions.

The study further notes, however, that *"Additional projects need to be studied and the database expanded for any further meaningful results to be drawn."*

The obvious question is how can this study vary so far from all of the other references to overtime and shift work that rather consistently predict a loss in productivity when additional hours are worked? It appears that the answer may well lie in the projects studied, the size of the database, and the analysis method employed. Some of the areas that must be questioned in this study are the nature of the projects, the method of measurement, and the skill of the study team.

In reviewing the material presented, the data was obtained only from seven projects. The data was collected by three students from the University of Texas working on their master's thesis. It is unknown if any of these students had any firsthand construction experience. It would seem appropriate that researchers involved in this area should have a reasonable level of experience in construction in order to be able to assess the validity of methods used and data recorded.

The seven projects studied were all of a process nature and included:

1. Refurbishing of a distillation unit at a refinery complex.
2. Expansion of a large oil refinery.
3. A shutdown of an oil refinery.
4. A power plant project.
5. Construction of a chemical processing plant.
6. A natural gas recovery plant.
7. A natural gas recovery plant.

In all seven projects, none were studied throughout the entire project but instead were assessed during limited time frames when overtime occurred. In some cases, the job was on overtime for the entire duration. In those cases, it seems highly questionable that any valid measurement could be made since a base standard was never established to define productivity without overtime. This leads to another important question about this study, that of the method of measurement.

The study did not measure like quantities but, instead, "normalized" different quantities to establish a baseline. For example, the study notes:

> *"In order to obtain useful periodic productivity values, it is necessary to equate the quantities installed to standard units. For instance, if a particular crew installs 6-inch pipe one day and 12-inch pipe the next, the two daily productivity values should vary significantly. Adjustment factors are clearly needed to equate the two commodities.*
>
> *The adjustment factors used in the CII study are known as "Bogey factors." Bogey factors are obtained by calculating the ratio of estimated work-hours to install one unit quantity of a particular material of detailed description with estimated work-hours to install the material of the size and description in question. For example, Bogey factors are used to convert all pipe sizes to 6-inch, carbon steel, field-run, socket-welded pipe."*

The conversion or normalization was done by graduate students in their master's thesis. Using estimates at any time in a study of this nature forces one to question the validity of the resultant findings.

When one reviews the detailed graphs presented in the study, it is obvious that the measured productivity varied significantly even for the same project over the time frame measured. Such variations indicate the lack of any pattern to the productivity achieved. This could be the result of many elements such as the measure used, the "Bogeys" calculated, the collection of information, etc.

Six years later, in 1994, the Construction Industry Institute (CII) issued another study titled: Effects of Scheduled Overtime on Labor Productivity: A Quantitative Analysis. The conclusions reached in this study were significantly different than the earlier study. The Executive Summary notes two major conclusions:

1. The first conclusion noted that curves shown in the Business Roundtable Report (BRT) were reasonable average estimates of productivity losses caused by overtime.
2. The second conclusion related to disruptions. The report noted three types of disruptions; Management disruptions, rework disruptions, and resource disruptions. The report

> concluded that "there is no clearly established relationship between the disruption frequency and the number of days worked per week" for management and rework disruptions. Resource disruptions were defined as the lack of materials, tools, equipment, or information. "*The frequency of resource disruptions increases sharply for longer workweeks.... The conclusion is that working overtime efficiently is a resource management problem.*"

Some other areas must be noted concerning this report. No longer were "Bogey Factors" used. Instead, the report utilized "Conversion Factors." In order to compare different types of installations, the report used an earned value approach. Since it was recognized that different size conduit, for instance, would require different time to install, the study defined a standard item for each category of installation. That standard item was assigned a conversion factor of 1.00. For different size conduit, for instance, any other size conduit was given a conversion factor compared to the 1.00 for the standard. The basis of these calculations was three estimating manuals and an estimating manual from one of the contractors working on one or more of the projects. The conversion factors were then used to determine how much of the standard item would have been installed. We will not enter into a discussion of the accuracy of estimating manuals. However, we will note that if ample information was available (and based on the report it was) a more realistic comparison could and should have been made of like items without using conversion factors. The use of conversion factors from a generalized estimating manual added subjective modifications to the analysis results.

The study incorporated data from four projects: a process plant, a manufacturing facility, a paper mill, and a refinery.[1] This clearly is a very limited sample size. Further limitations were that the "focus" of the study "was on detailed observations of piping and electrical crews, rather than on various crafts." Yet, electrical observations were only made on three of the four projects (process plant, manufacturing, and paper mill) and mechanical observations were also only made on three of the four projects (process plant, manufacturing, and refinery). The obvious question is whether the data can be applied to drywall work, concrete work, etc. The application to other trades has not been substantiated by the study.

---

[1]See Table 1 within the CII SD-98 (August 1994).

Furthermore, the study defined a normal workweek as four 10-hour days. Overtime was then defined as five or six 10-hour days. "There were no data for 5–8-hour days."[2] Once again, the obvious question is the applicability of four 10-hour days since many construction projects operate on the basis of five 8-hour days or five 10-hour days.

The hours used in the study were also modified from the raw data. As stated in the study, "If the crew worked 37.5 hours during the week, that was considered a 40-hour week."[3] It is unclear why the study would not just use the raw data of a 37.5-hour week. As explained, the additional 2.5 hours in the conversion would not *appear* to be a significant change, but not all conversions were completed in a like manner. The samples for "Actual Hours" and "Nominal Hours," the hours used in the analysis, were provided in Appendix A, Table A-1. For Project 9187, Week 2, 33 Actual Hours were reported, and the study converted those hours to 40 Nominal Hours, resulting in an addition of seven nonproductive hours. In this example, the quantities completed during this week would have seven additional hours added, showing a lower productivity measurement than what was actually achieved using 33 hours. One would think given this example, any time the Actual Hours equaled 33 then the study rounded up to 40 Nominal Hours for consistency reasons. This was not true.

Project 9188, Week 4 also had 33 Actual Hours, but in this instance, the study rounded down to 30 Nominal Hours. When compared to the other sample where 33 hours were increased to 40, the sample that reduced the hours from 33 to 30 would show 10 fewer hours used to achieve the quantities obtained during the week (40-30=10). The 10 hour difference resulted in a modified productivity rate because of the subjective changes to weekly hours.

In two other examples, 42.25 Actual Hours were rounded up to 50 Nominal Hours (Project 9186, Week 9) while 35.5 Actual Hours were rounded down to 30 Nominal Hours (Project 9189, Week 4). Of the 132 data samples for hours, 43 samples (over 32.5%) had the Actual Hours modified to another quantity of Nominal Hours. Stated differently, over 32.5% of the productivity calculations were based on hours that did not actually exist in the raw data.

The study also did not identify which weeks were used to calculate the "Baseline Productivity," which appears to be the alleged productivity

[2] See page 30 within the CII SD-98 (August 1994).
[3] See page 20 within the CII SD-98 (August 1994).

achieved during non-overtime periods. The study stated that, "The weeks used for the baseline are noted in Appendix A by an asterisk (*)." However, no asterisks were shown in the version of Appendix A, purchased from CII.

Other data "outliers" were also excluded from the analysis. Examples of raw data collected, categorized as "outliers," and excluded were:

- The first week of data for three projects was excluded. These exclusions were attributed to "some initial difficulties with data collection....for projects 9181, 9183, and 9185.[4]
- One 7-day workweek was discarded.[5]
- All performance factors greater than 2, which would show higher levels of productivity, were "ignored".[6] This appeared to occur on 11 samples.[7] The specific reason provided for these exclusions were:[8]

> *"The analysis found that the results were very sensitive to extreme values. Therefore, all PF [performance factor] values greater than two were ignored."*

The alleged "extreme values" were excluded despite the study also stating:

> *"However, simply removing extreme data points would be improper since they are, to some extent, the focus of this study."*

The analyses and conclusions were completed once the non-project-specific conversion rates were used to "normalize" the work comparison, the alteration of productivities were changed by subjectively modifying the "Actual Hours" to "Nominal Hours," and the excluded data showing higher productivity results were excluded. Similar to our advice concerning the study performed by the CII six years earlier, based on the preceding issues, the use of this study in any form is questionable.

---

[4] See page 21 within the CII SD-98 (August 1994).
[5] See page 23 within the CII SD-98 (August 1994).
[6] See page 23 within the CII SD-98 (August 1994).
[7] See Appendix B, Table B-1 within the CII SD-98 (August 1994). Note that the quantity of samples in Table B-1 is different than the samples in Table A-1.
[8] See page 23 within the CII SD-98 (August 1994).

Quantitative Impacts of Project Change

A study issued by the CII in May 1995 titled "Quantitative Impacts of Project Change" addressed changes on craftsmen productivity. The study was based on questionnaires sent to 90 CII member companies. Responses were received from 35 different companies and represented 104 projects. It should be noted that almost half of the responses were from owners as opposed to contractors or construction managers. This gives rise to the question of the accuracy of productivity data available. The projects consisted of manufacturing and process work. There did not appear to be any vertical or heavy civil projects included. The measurement of productivity was based on a ratio of earned hours to expended hours. Numerous statistical evaluations were performed on the data, and the report attempted to answer three hypotheses concerning changes.

The findings of the report can be summarized as follows:

> Hypothesis 1: *"Changes that occur late in a project are implemented less efficiently than changes that occur early in the project."* The report concluded that this hypothesis was valid but could not statistically support it.
> Hypothesis 2: *"The more change there is on a project, the more of a negative impact there is on labor productivity."* The report concluded that this hypothesis was validated statistically.
> Hypothesis 3: *"Hidden change increases with more project change."* The report concluded that this was statistically supported.

While the study supports hypotheses which have been recognized for quite some time, there does not appear to be any quantifiable result from the report which can be used in assigning measurable losses of productivity because of specific amounts of changes. Furthermore, because the data upon which the study was based is the result of questionnaires and not measured firsthand by the researchers, one must question the reliability, accuracy, and validity of the source data. Finally, because the data is from limited types of projects, and is based on an "earned value" approach, the results must be questioned as to the accuracy of the information. It should be noted that on page 3 of the study, it states:

"Because the results indicate relationships from a broad, total project perspective they cannot be used to accurately calculate the cost of an individual change or a group of changes. The findings of the study are helpful in benchmarking the amount of change and the expected productivity on a given project against the studied projects. However, many variables affect productivity on a given project other than the amount of change, such as the timing of changes, the rate of changes, the change management system in use, and the type of changes. In evaluating the productivity on an individual project, these other variables, whose effects were not studied in this research, must be taken into account."

The Ibbs Studies

Dr. C. William Ibbs is a professor at the University of California at Berkeley.[9] Dr. Ibbs has authored multiple publications on various topics in the construction industry. The Ibbs studies are another type of industry study that allegedly support those changes that equate to productivity and project losses.

For a bit of background on the Ibbs studies, Dr. Ibbs and Mr. W.E. Allen co-authored "Quantitative Impacts of Project Change" in May 1995 with the Construction Industry Institute at the University of Texas, in Austin, Texas.[10] The 1995 Ibbs/Allen study included 104 projects that compared the percentage of change orders (the independent variable) to a productivity factor (the dependent variable). Percentage of change orders was determined by the number of labor hours expended on authorized changes during the construction phase divided by the total labor hours expended during the construction phase. Construction Productivity Ratio, or Index,[11] was measured by taking actual productivity divided by planned productivity. The following Figure 8.9 is from the 1995 Ibbs/Allen study and shows the 104 projects plotted to show the percentage change orders compared to the productivity index:[12]

---

[9] List of publications by Dr. Ibbs on University of California at Berkeley website, accessed May 2, 2021: https://ce.berkeley.edu/people/faculty/ibbs/publications

[10] Ibbs, C.W. and Allen, W.E., "Quantitative Impacts of Project Change," Construction Industry Institute, Source Document 108 to Publication 43-2, May 1995.

[11] As acknowledged in the 2005 study by Dr. Ibbs, it was assumed that the "planned productivity" was accurate and could be used as a comparable basis to determine productivity loss.

[12] McEniry, Gerald. "The Cumulative Effect of Change Orders of Labour Productivity – the Leonard Study "Reloaded," *The Revay Report*, Volume 26, Number 1, May 2007, Figure 4.

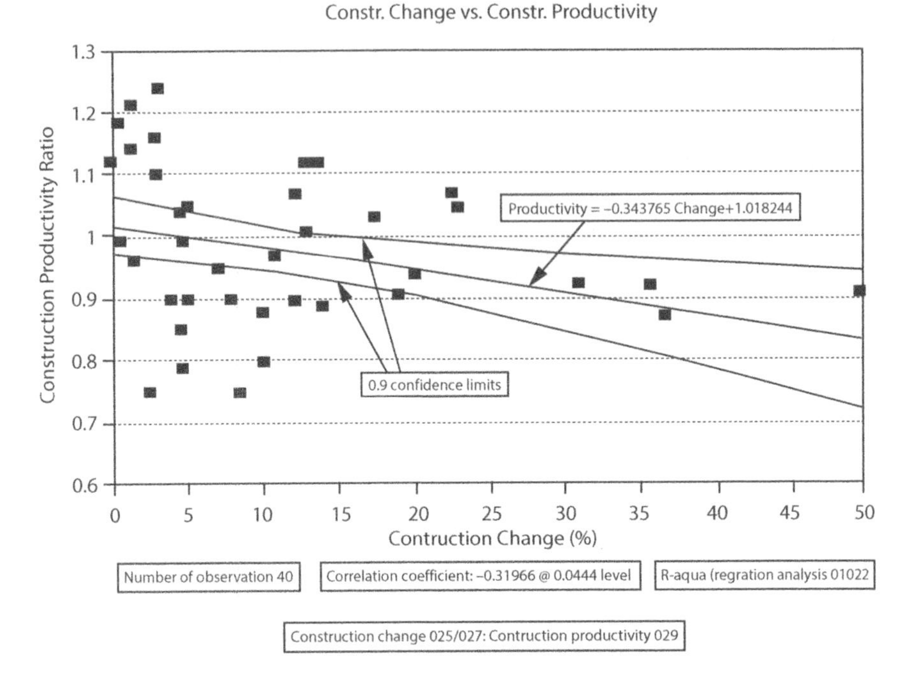

Figure 8.9 1995 Ibbs/Allen study.

Some general observations on the information in the 1995 Ibbs/Allen study:

1. Most of the data exists in the 0% to 20% range for % Change Orders. There was little data provided where the change orders exceeded 20%.
2. The data within the 0% to 20% Change Orders range has a high amount of variability.
   For example, projects in the 10% to 15% Change Orders range appear to have the same frequency of greater than 1.00 productivity (better than planned) instead of lower than 1.00 (productivity losses against the plan). This, along with most of the data points from 0% to 20% Change Orders, would indicate that the variation in the productivity index is not dependent on the variation in the percentage change orders.
3. The best-fit, or regression model, is linear and has a weak coefficient of determination (R^2) of 0.1822.

In simple terms, an R^2 of 0.1822 means that 18.22% of the variation in productivity may be explained by the variation in change percentage. In other words, even if a causal relationship actually existed between the two variables, 81.78% of the productivity variation remained unexplained by a simple regression comparison between the two variables.

In 2005, Dr. Ibbs authored an update to the 1995 Ibbs/Allen study and included the data from 169 projects (65 more projects than the 1995 study). The 2005 publication,[13] which was titled "Impact of Change's Timing on Labor Productivity," supplemented the data from the 1995 publication, and further classified productivity impacts into three categories based on the timing of the changes (early, normal, and late). The following Figure 8.10 shows the Ibbs study data for the 2005 analysis plotted similarly to the 1995 data:[14]

Several observations from the preceding 2005 Ibbs study figure are:

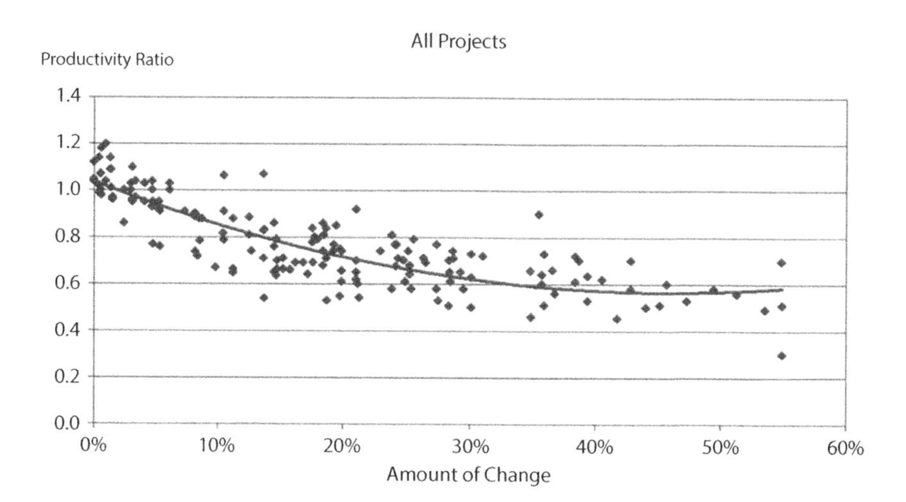

Figure 8.10 2005 Ibbs study.

[13] Ibbs, William, "Impact of Change's Timing on Labor Productivity," American Society of Civil Engineers, Journal of Construction Engineering and Management, November 2005, pages 1219 through 1223.

[14] Ibbs, William and Vaughan, Caroline, *Change and the Loss of Productivity in Construction: A Field Guide*, Version Date: February 2015, page 68. Source data link: https://ibbsconsulting.com/wp-content/uploads/2020/03/Change-and-the-Loss-of-Productivity-in-Construction-A-Field-Guide.pdf

1. The 2005 Ibbs study appeared to include more data in greater than 20% Change Orders range that was lacking in the 1995 Ibbs/Allen study.
2. There were multiple projects in the 15% to 25% Change Orders range with a productivity ratio greater than 1.00 in the 1995 study that are not shown in the 2005 study.
3. There appears to be reduced variability in the 2005 Ibbs study. This would appear to be an improvement compared to the 1995 Ibbs/Allen study.
4. The best-fit, or regression model, has been changed from a linear 1st-order equation to a 2nd-order equation.
   The regression model contained the following equation:

$$y = 2.4621x^2 - 2.169X + 1.0589,$$

where x = Amount of Change and y = Productivity Ratio

   Using this equation in attempting to describe the relationship between change and productivity is problematic, as will be discussed later in this section.
5. Coefficient of Determination (R^2) has improved to 0.72, indicating that a higher percentage of variation in the Productivity Index (y) may be explained by the variation in the Amount of Change (x).
   A stronger R^2 would also appear to be an improvement compared to the 1995 Ibbs/Allen study.

Dr. Ibbs's 2008 article,[15] "Evaluating the Cumulative Impact of Changes on Labor Productivity – an Evolving Discussion," provided a graphic that compared the results and data of the 1995 and 2005

[15] Ibbs, William and McEniry, Gerald, "Evaluating the Cumulative Impact of Changes on Labor Productivity – an Evolving Discussion," AACE's *Cost Engineering*, Volume 50, Number 12, December 12, 2008, pages 23 through 29.

information. The graphic shown in Figure 8.11 is the version included in Dr. Ibbs's 2008 article:

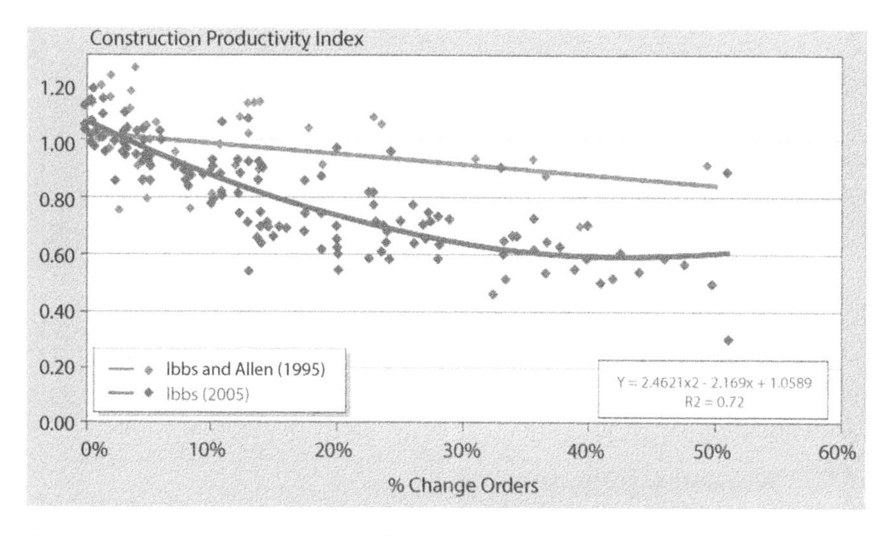

Figure 8.11 Comparison of 1995 Ibbs/Allen study and 2005 Ibbs study.

The preceding figure allows for a comparison of the 1995 Ibbs/Allen study and 2005 Ibbs study and provides several key observations:

1. Data that was included in the 1995 Ibbs/Allen study was omitted from the 2005 Ibbs study.

   Many of the 1995 data points (grey) that showed a higher productivity when compared to the 2005 data points (red) were removed. Removing the 1995 values would reduce the variation among the data. Why the 1995 data points were omitted was not addressed in either Ibbs' 2005 or 2008 publications on this topic.
2. The polynomial equation used in the 2005 Ibbs study shows that when change order percentages exceed 44%, the productivity index will improve. In other words, the more the change orders exceed 44%, the better the productivity will become.

This conclusion is counter to many of the assertions in the 2005 Ibbs study. This issue can be observed by the inversion of the regression line after 44%, and how it bends upward after 44%. To further illustrate this point, had the 2005 Ibbs study included change order values out to 100%, the 2005 Ibbs study regression model would support a productivity index of over 1.3, as shown in the chart in Figure 8.12.

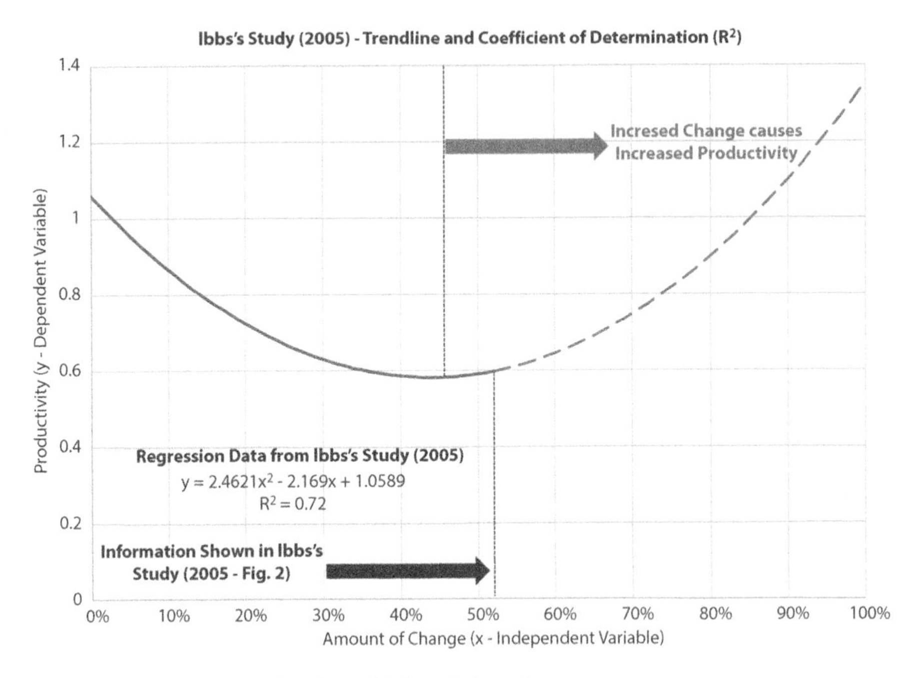

Figure 8.12 Regression data from Ibbs' study (2005).

This begs the question—is the model only accurate up to 44% and, if so, how could it be used to reasonably determine how percentage change effects productivity with such a significant limitation?

If the regression model equation is correct then the 2005 Ibbs study would support that if a project incurred 87% change, then the productivity index would be 1.035—a productivity improvement compared to a 1.0. The following use of Ibbs' 2005 regression equation supports this alleged outcome:

$$y = 2.4621x^2 - 2.169X + 1.0589$$

$$\text{If } x \text{ (Amount of Change)} = 0.87\text{, then:}$$

$$Y \text{ (Productivity Index)} = 2.4621(0.87)^2 - 2.169(0.87) + 1.0589 = 1.035$$

Increase the change percentage more, and the productivity index improves even more. If the equation were true, a change percentage of

200% would result in a productivity index of 6.569—a massive productivity gain to the project. The following use of Ibbs' 2005 regression equation supports this alleged outcome:

$$y = 2.4621x^2 - 2.169X + 1.0589$$

$$\text{If x (Amount of Change)} = 2.00\text{, then:}$$

$$Y = 2.4621(2.00)^2 - 2.169(2.00) + 1.0589 = 6.5693$$

It gets worse from there, but you get the point—the equation in the Ibbs 2005 study cannot be used to accurately measure the amount of productivity loss caused by change.

The 2005 Ibbs study concluded that, "The research reported in this paper (and other studies) reaffirms that project change is disruptive to labor productivity." Simply put, this conclusion is not supported by Dr. Ibbs' regression model in the 2005 Ibbs study. For resolution of doubt, the authors of this book requested a copy of the underlying data used in Ibbs' studies in 2013 but were informed that the information was "proprietary to [Dr. Ibbs'] consulting practice."

To our knowledge, the 2005 Ibbs study was the latest version of the Ibbs study, and it has not been revised since. Dr. Ibbs' publication[16] from 2015 also used the 2005 Ibbs study information and did not contain an updated study. Ibbs' 2018 publication stated:

> *"We hope this field guide gives you the information you need to gain perspective on what change is and the possible effects it can have. Keep in mind that all of the methods presented are generalized and it is important to remember that every job is unique and will run into its own problems. Many of the methods and factors described will have slightly different effects on each job, and it is necessary to incorporate the unique characteristics of your project when estimating the effects of change. (Emphasis added)*
>
> *We cannot stress too much that the guidelines presented are approximate."*

---

[16] Ibbs, William and Vaughan, Caroline, *Change and the Loss of Productivity in Construction: A Field Guide*, Version Date: February 2015, page 5. Source data link: https://ibbsconsulting.com/wp-content/uploads/2020/03/Change-and-the-Loss-of-Productivity-in-Construction-A-Field-Guide.pdf

The authors support Dr. Ibbs' statement that "every job is unique." However, we must emphasize that the analysis of the unique characteristics that exist within each job are essential to determining the productivity losses that occurred on that job. Every project, its design, and its requirements are unique. Every contract is unique. The contract requirements are unique. The applicable law is unique. The people administering the contract are unique. The change administration requirements are unique. The people administering the changes are unique. The change estimates are unique. The people performing the change estimates are unique. The people tracking the productivity information are unique. The tracking systems and capabilities of those systems are unique. The negotiation tactics are unique because the people employing those negotiation tactics are unique. The desire to settle, financial and operational capabilities of the parties, and desired outcomes are unique. We could keep going, but you get the point. These are just several examples of the unique aspects that exist between every project. As such, it would be imprudent to trust that any "study" could reasonably opine on the unique characteristics of any project.

The criticisms of these studies are not limited to this book. Similar sentiment was offered by Mr. Gerald McEniry in an article from 2007.[17] Mr. McEniry's article included his evaluation of the Leonard study and Ibbs' study (versions dated 1995 and 2005), and while we do not agree with all the opinions within the article, we agree with the following statements from Mr. McEniry's article:

> *"Generic industry studies should never replace project-specific information....*
>
> *Lost productivity is best studied by evaluating causes and effects specific to a particular project...Industry studies alone are of limited use."*

We could cite multiple sources that discourage, if not denounce, the use of industry studies, but this would belabor the main point.

The hard work of documenting and analyzing these specific circumstances for each project should not, cannot, be neglected or bypassed by the analyst. It may be more challenging to analyze project data that is

[17] McEniry, Gerald. "The Cumulative Effect of Change Orders of Labour Productivity – the Leonard Study "Reloaded." *The Revay Report*, Volume 26, Number 1, May 2007, page 8.

voluminous or contain imperfections. However, it would be hypocritical to then suggest that a "study" with even less commonality to the subject project could reasonably determine what happened on the project. That, along with the other reasons presented, are why these "studies" are inappropriate as a measurement tool or standard. These statements are not meant to demean the authors of these "studies," or the time expended to produce them. Rather, the industry must continue to move toward understanding the process to track this information, the purpose and value it holds, and then hold analysts accountable for using that data as a source for project analysis.

Other Reports and Studies

There are numerous other related reports and studies. Many of these are included in the bibliography of this book. While it might be intellectually pleasing to discuss every one of these studies, it is not believed to be productive. The studies and reports noted previously appear to be the more prevalent ones that are often referenced in the industry.

Use of "Industry Publications"

The overriding question concerning these industry publications is how do we use them? Clearly, none are definitive enough or have enough scientific bases to be used as a direct measure of loss of productivity on a construction job. To attempt to solely utilize these publications as a quantification of lost productivity would be incorrect. Rather, these publications can be used as support for a more direct study of actual project data such as a measured mile analysis. Similarly, they may be used in conjunction with expert analysis and opinion as support for those expert conclusions.

9

COVID and Productivity

Introduction to COVID Impacts

As we were approaching the completion of this book in early 2020, the world was hit by the COVID pandemic. This widespread event affected operations throughout the construction industry. All facets of the supply chain were affected. On-site labor was affected. Project management and staffing was affected, either having to divert resource toward managing new COVID-related issues or figuring out how to have some of the staff manage work remotely. In general, there were few areas of project operations that were not affected in some capacity. This chapter will address some of the effects that COVID had on construction productivity, and how this unique circumstance may alter the approach to measurement and recovery of losses incurred because of COVID inefficiency losses.

COVID's Effect on Construction Projects

The COVID pandemic was a "perfect storm" to construction productivity. It introduced new challenges via supply chain and production shortages, compounded existing concerns with craft labor shortages for ongoing and upcoming projects, and in some instances, added new management and resource complexities because of the spike in remote, or virtual, work completed on projects. This section will describe *some* of the impacts incurred on projects during the COVID pandemic.[1]

[1] Note that the authors started this chapter at the beginning of the COVID pandemic. The book was submitted for publication in early 2022 when the COVID pandemic, its many variants, and the changing effects of the many variants were still ongoing. We anticipate that the ramifications and outcomes of the COVID pandemic will continue to evolve and, as such, so will the applicable data on how COVID affected construction projects and productivity.

Ted Trauner, Chris Kay and Brian Furniss. Productivity in Construction Projects, (143–158) © 2022 Scrivener Publishing LLC

Lumber Supply Chain Example

As an illustration, we'll provide an example of how the lumber industry was affected by COVID. Some examples of effects on lumber milling operations were:

- Reduced working times at the mill due to imposed curfews and lockdowns.
- Increased spacing and testing requirements to enter the mill to start productive work.
- Reduced productivity for workers in lumber mills as newly imposed safety requirements, including increased spacing and testing, reduced the overall output and production of the mills.
- Milled lumber crossing provincial or national boundaries were affected by the changed laws of countries that either required increased testing or, in some instances, completely shut down the borders to importing and exporting operations.
- This lowered the available lumber stock and created material shortages at lumber yards.
- The material shortages caused lumber prices to increase substantially, resulting in increased material costs due to supply and demand constraints.
- Some transient workers housed in camps near lumber harvesting facilities and mills were either affected by COVID impacts at their jobsite or reduced need at the mills because reduced output caused a raw lumber supply build-up at the mill.
- Even when lumber was available at an increased cost, the worker productivity and production at sites often also decreased because of the increased safety standards and monitoring at the site due to new COVID protocols.

There are more examples, but you get the point. From the beginning to the end of the supply chain, various aspects were affected by COVID, causing massive disruptions and price fluctuations.

Impacts on Global Commerce and Resources

The examples of COVID impacts are not limited to the lumber industry, of course. In the United States, materials continued to flow, but the already-existing dependence on long-haul commerce, materials, and labor mobility had a tremendous effect on overall plan execution, the workers available to execute that plan, and the timing and costs of materials. International projects were hit even harder because procurement, management, and field labor was even more dependent on international working crews, commerce, and mobility, and these areas suffered even larger effects than domestic projects. For example, it's more challenging to test and certify operation on a valve already fitted on a Middle Eastern project when the valve manufacturer is in Europe, and the specialty crew that performs that testing is precluded from leaving their home country or entering the country where the project is located. With the worker in one country and the equipment in another country, the worker will not be able to fully observe and appreciate the "hands on" performance of the equipment during the testing process. How confident is the worker in his/her ability to start and utilize that equipment? What effect will that have on warranty cost, coverage and replacement?

Many companies developed new and innovative means attempting to mitigate the effects of the issues. While those efforts may have reduced the effects of a total time or productivity loss, or simply prevented them from being worse, they often did not eliminate the effects on the projects.

Additional Effects to Craft Labor Shortages

Construction projects, and the companies performing construction work, were already encountering difficulties finding adequate skilled craft labor to perform the field work *before* the COVID pandemic period. Craft labor challenges were not new to the industry and occurred in previous decades when the construction industry felt the ebbs and flows of economic changes. Before COVID, the United States had entered a time period where a large number of highly experienced craft labor and management were either retiring or nearing retirement. The resource pool required to supplement and replace the retiring workforce was not sufficient, and companies were forced into a situation where it was increasingly harder to find available resources adequately skilled to complete the work. As is typical in the supply-demand curve, labor supply was reduced and the cost to procure the limited skilled resources increased via escalating wages and benefits for craft labor. This placed existing and future projects in a bind and created a competitive craft labor "war" between construction companies. Then COVID happened.

COVID took an already challenging labor market and added more chaos. Now those limited resources had to be further limited due to COVID safety protocols. Labor resources that traveled via plane to complete projects stopped traveling due to COVID concerns or shifted to other travel means that decreased the time available to work on the projects. For example, imagine a project manager completing a project in Boise, Idaho, while her family was located in Seattle, Washington. Before COVID, the project manager went on a plane on Sunday evening from Seattle to Idaho, performed work during the normal Monday through Friday working cycle, and departed the site on Friday to return home to her family over the weekend. With COVID, the project manager's travel to the project site was reduced due to personal medical concerns, family safety concerns, new company safety protocols, or other reasons that came out of the COVID pandemic. The project manager began to travel from home to and from the site in the company vehicle, turning the former three-hour plane travel into an eight-hour trip on Sundays and Fridays. This caused the project manager to spend even less time with her family and added increased personal issues. This continued until the project manager started performing more work remotely. While technology allowed for some work to successfully occur remotely, there are situations when there is no substitute for solving the challenges at the project site. The situation is further compounded when the team supporting the project manager is also fractured into remote working and is facing its own COVID-related challenges.

Problems affecting field craft labor were even worse because 'remote working' was not an option for craft labor. The craft work faced similar travel and personal challenges as the project manager, but the craft worker had to be at the site and around other people in order to get the work accomplished. For union jobs, skilled labor that was already in high demand pre-COVID increased further, and a "bench" of lower skilled workers was no longer available. Of course, productivity may be adversely affected when a less-skilled labor crew is the only resource available. Consider the negative effects on construction productivity when less skilled workers were forced to take on a larger role and working crews that required close contact were now limited to smaller sizes to comply with company or jurisdictional safety requirements. When you combine these COVID-specific impacts with the pre-COVID labor "war" that was already ongoing, most construction projects suffered increased labor productivity losses.

We could continue trying to present the complexities that labor resources faced during the COVID pandemic, but we would still not cover *all* the specific impacts encountered on the projects. The COVID pandemic layered so many resource issues into our personal and professional lives, and these issues caused craft labor and management labor effects that hindered most projects.

Evaluation of Productivity Improvements Due to COVID

We would be remiss if we did not mention that there may have been situations when productivity *increased* due to COVID. One could imagine a situation when reduced highway traffic created fewer lane closures, or reduced airport traffic opened more runways, taxiways, and aprons for work and reduced the traffic anticipated and actually incurred in the pre-COVID time period. For example, imagine a roadway project where lane closures were restricted to times when traffic was in a reduced-flow state. COVID resulted in fewer people traveling on roadways and, therefore, the reduced-flow state of traffic occurred more often or became the norm for a time period. In certain circumstances, this allowed transportation departments to reduce impediments for ongoing construction projects and shifted what were previously non-working periods to be available work periods. Likewise, reduced airport traffic meant fewer operating gates. Fewer operating gates meant reduced traffic on aprons, taxiways, and tarmacs. In these circumstances, you could see an airport authority reducing the areas used for aircraft transport into and out of the runway to reduce the staff needed to operate and maintain the airport. Airport authorities may have made these areas open to construction planned during COVID for longer periods than originally anticipated, as performing construction in these areas no longer impeded airport operations.

You could see where this could benefit contractors operating under normal labor circumstances. However, these anticipated improvements may have also been offset by the COVID labor and safety impacts previously described. For example, it would be an apparent benefit to have less airplane traffic and two aprons available for a contractor to perform resurfacing operations simultaneously in two different areas. However, the contractor would need two crews instead of one crew to perform the work simultaneously. Given the COVID labor impacts and safety protocols, the contractor may not have been able to procure a second paving crew, and the productivity of the original paving crew may also be adversely affected. The point is, there may have been circumstances when productivity could have increased but due to other COVID impediments, those anticipated productivity increases may have been reduced, if not eliminated. This highlights the importance of each contractor recording the effects on their work, along with the actual performance of quantity information and hours expended during the time periods to allow for the monitoring and measurement of productivity effects.

Material Price Fluctuations

COVID caused dramatic swings in the supply chain and resulted in price fluctuations. Steel, concrete, fuel, lumber, and labor pricing all fluctuated with increased variability during COVID. For example, at the beginning of the pandemic, fuel prices were reduced as more people worked remotely and reduced the demand for fuel. However, this was only a temporary drop. By early 2021, fuel prices had increased approximately 50% compared to the pre-COVID fuel prices as people returned to the roadways and global supply chain concerns continued. Certain contracts allowed for evaluation and recovery of commodity price fluctuations, but many hard-bid and guaranteed maximum prices do not allow for recovery of commodity price increases.

Contractors during the COVID period and other particularly volatile time periods may benefit from negotiating commodity price corrections in their contract. Owners may perceive such risks to be an unnecessary burden in contracts, but it may actually reduce the bid pricing received by owners because the potential risk to the contractors is reduced and, as a result, contractors can carry less money in the bid to cover the potential risk. In short, it may be beneficial for parties to reconsider and revise their risk profile and agree to adjustment procedures to handle such price fluctuations that cannot be reasonably predicted by any party.

Entitlement to Recover COVID Productivity Losses

To our knowledge, owners, contractors, suppliers, and various parties are openly acknowledging that COVID has caused increased impacts. What the parties are struggling with is how to resolve who is responsible to pay for these impacts, as few contracts addressed the risk of a pandemic and whether parties are entitled to the additional time and costs caused by COVID. We have observed various legal arguments that COVID impacts are force majeure, or "superior force" or "act of God," impacts. The party at risk for the time and cost aspects of force majeure impacts varies by contract. Contracts often provide the contractor excusable time for force majeure impacts, but not compensation for those impacts. This determination would seemingly force contractors to absorb the additional costs incurred, despite the causal events being outside of their control. Owners would equally absorb the effects of project delays due to COVID, having to absorb the additional cost of project oversight, financing, lost revenue, and other project time-related costs to owners. Being as this is a book on productivity, some productivity losses may cause delay and, as a result, the productivity losses may be subject to the time-related aspects of force majeure clauses.

Some legal counsel have also asserted that certain impacts of COVID were caused by changes in law, either on the national, state or province, and local level. For example, recall the prior example about testing and certification on a Middle Eastern project. Presume the equipment operator that performs the tests on behalf of the equipment manufacturer cannot leave his home country of Germany due to a change in German law. Perhaps German law allowed the travel of that worker, but the United Arab Emirates (UAE), the country where the project is located, precluded travelers from Germany from entering the UAE. Another situation may be that the worker could travel from Germany to the UAE, but while he is in the UAE, laws changed in either Germany or the UAE and precluded the worker from leaving the UAE. Now, presume the contract addresses changes in law differently than force majeure, and allowed the contractor to recover for additional time and costs caused by changes in law. In this situation, the contractor would not only be able to recover the additional time (similar to the force majeure example), but he may also be able to recover additional costs resulting from changes in law (different than the force majeure example). This would be a large difference in potential recovery depending on what issues caused which effects.

Again, increased costs resulting from productivity losses incurred due to changes in law would be recoverable in this example, along with the delays that resulted from productivity losses.

Who would pay for the additional lost time, travel, and other costs for having the worker stuck in a foreign country with no ability to return home? The risks of how to safely transport workers to and from the project sites has always been a consideration of companies, especially those companies performing work internationally, but the potential of various jurisdictional changes on such a wide-spread event is "new territory." This example not only highlights the entitlement challenges presented to projects and legal counsel, but it also highlights the importance of tracking and segregating impacts in a detailed manner. The contemporaneous recording of impacts and costs, and whether those items were caused by force majeure events or changes in law, may have significant effects on owners and contractors throughout the industry. We encourage the parties to seek guidance from competent legal counsel on their potential means of risk and recovery relating to COVID.

Beyond the questions of foreign travel and construction, there were just as many variables in the United States, and those variables changed whether it was the original COVID, the Delta variant, or the more contagious but less deadly Omicron virus. But in each case, there was a delay claim. How were these addressed?

The authors are familiar with companies that dealt with hundreds of delay claims where productivity suffered, all of which were analyzed along the following lines:

(a) The nature of the job;
(b) Where the job was in the construction process when the COVID-related delay first occurred and when it passed;
(c) The nature of the labor force;
(d) The nature of the party's compliance protocols; and
(e) The level of government involvement.

As to the nature of the job, great attention should be directed to the unique aspects of the project itself. Was this the construction of an open-air stadium or even an arena where the work was still performed outside until the roof was erected? Was this project the construction of an office building, or some other vertical building, where social distancing had a far greater impact in the construction process? Or was there a need to change the design of the building to meet new social distancing protocols, such as going from an open floor format to offices? Asked more generally, what was the air circulation on the job site and intended for the completed facility?

The progress on the job at the time it was shut down or adversely affected by COVID is also important. For example, was the steel partially or fully installed? If it was only partially installed, was the rest of the steel on site? If not, was it delivered in a timely manner, or was it subject to the supply chain delays that manifested themselves a few months later? Or was the delay instead related to the Omicron virus, which tended to wipe out large amounts of the labor force on a given job site—but with those losses only present for a few weeks. How quickly did the labor force return to work, and in what areas?

The nature of the labor force differed remarkably by region of the country. In those areas where there are high hourly rates, either due to union contracts or "prevailing rates," many workers returned to work as soon as they were permitted to do so to earn those higher wages. In those areas of the U.S. where the government shutdowns were relatively brief, workers also tended to return to work after the brief hiatus. High-profile or unique projects, the kind that generate pride amongst the workforce and the community, also tended to have a greater and quicker return to work.

But there was evidence in some jobs where certain types of workers took full advantage of the ability to collect government checks and not work. Others would take advantage of the fact that one of their fellow workers tested positive, and they used the protocols to stay home for several days, despite no symptoms or subsequent illness. That reluctance to work, or to take advantage of the opportunity to get paid without working, not only made it more difficult to get people back to work, but also made it very difficult for contractors to "catch up" by going to a full second shift. Despite the desire to do so, in many instances there simply were not enough workers—or supervisors. In other cases, contractors could not even initiate an overtime program, or only operated it on a partial basis. Saturday work was compromised, due to a lack of labor or the fact the workers were physically and/or mentally fatigued.

The nature of the workforce affected by COVID was also different in another important context: who was affected? A contractor could have a full complement of his workers ready to go once the architects or engineers gave their approval. But those architects or engineers might have been out for weeks battling COVID. By contrast, the architects and engineers could be fully engaged, but Omicron wiped out a substantial number of members of a given group of trades, such as the electricians, for several weeks.

A contractor's ability to quickly institute an efficient compliance program was also critical. With 1,200 people on the job site, the ability to convey the COVID requirements in five of fewer minutes, or to have a crew come in after the day shift to clean the entire job site, along with equipment and tools, created an opportunity for some contractors to get greater productivity out of their highest paid and most critical workers. Comprehensive contact tracing programs could limit the number of people that took time off for precautionary reasons. In some instances, well-documented compliance programs could generate greater trust in the community and with the workforce, thus increasing the productivity of the workforce and decreasing the time spent on local government compliance issues.

As to the level of the government's involvement, the manner in which various states handled the shutdown, and the length and breadth of that shutdown, is well documented. Some jobs, like the cities in which they were located, were completely shut down. But that was not always the case. Somewhat counterintuitively, some contractors did very well in states enduring long shutdowns, precisely because all the generally smaller projects were closed down. In those instances, sometimes for only a limited period of time, the labor supply greatly exceeded the demand for those workers. By contrast, contractors in other states discovered that no one was willing to take a cut in pay to go back to work because jobs were so plentiful. But even in some of the states that had the most restrictive response to COVID, there were instances where the local governments were either helpful in getting construction projects moving again or did not provide active COVID oversight once the work resumed. In those situations, the greater challenge for the contractors was to get the public at large to understand that there were good reasons to return to work.

In addition to these five factors, the authors also note the "problem-solving spirit" of some contractors and construction managers. Some contractors would lengthen the workday as part of a deal they negotiated with the workforce. Others sought to work with the owners to extend the finish date, with documentation to justify the belief in that new finish date. Others came up with temporary solutions to accelerate the opening. For example, they may have ordered several hundred toilets for the facility, but with supply chain issues they could not get all of those from the same vendor. Instead, they installed a different type of toilet for the opening, and then determined later with the owner whether they should replace them with the same toilet as originally intended.

Measuring COVID Productivity Losses

It may be surprising that despite COVID effects incurred on so many levels, there is no "one size fits all" ruler to measure the effects on each job. For example, a superstructure crew building a highway and bridge project in Texas would have fewer effects if it had already procured lumber and rebar for forms and reinforcement and executed subcontracts with the subcontractors preparing to perform the work. This is not to suggest that the prior example would not have encountered some effects to its operations and costs, but it often incurred fewer impacts than the company in the earlier stages of subcontractor and material procurement.

Given the variable effects on each project, how would a management team measure the productivity losses? As you may have guessed, some industry-based companies and organizations attempted to create new "studies" that attempted to generalize and quantify certain effects of COVID. The raw data in these "studies" were limited, if available at all. The lack of raw data may be understandable given how the recent nature and magnitude of effects may limit the amount of data available, but that lack of data must also be considered within the context of its application and usefulness to measure productivity losses. Further limiting the use of these studies was how the effects varied from project to project. Each country created different policies and enacted varying changes in law. The same goes for each state or province. Corporations created different policies, and the tasks and enforcement requirements varied depending on location. You get the point—the impact effects vary widely from project-to-project. If a supply chain was dependent on material or labor production in China, it incurred different impacts than the project solely dependent on materials produced within its same country. The same may be true for two similar projects with the same company within the same area, but in different phases of the project.

Moreover, the overlap of COVID effects further complicates the relationship between causal factors and productivity losses. Earlier in the book, we discussed several simple examples using linear regression analysis to aide in causal determination. The examples used pinpointed the effect of a singular event, or single factor, that affected the dependent variable (productivity). With COVID, the project may be incurring multiple issues, or multiple factors, and the timing and magnitude of effects from those factors may be constant or variable throughout the project.

We'll save you the suspense—our recommendations are largely the same as they were throughout the rest of the book. Without a "magic bullet" or reputable study that can accurately quantify the effects of COVID on each specific project, the solution to document COVID issues presented an overwhelming challenge to project management. Project management either had to create new teams to document and measure these impacts, or it had to sharpen its approach to track these impacts with the resources available.

Recommendations for Measuring Productivity Losses during COVID

The following subsections discuss methods of measuring productivity losses that were presented, in detail, in other areas of this book. The deployment of these methods will not be restated in this section, but certain COVID-related considerations of the methods will be emphasized.

The Measured Mile and "A Day in the Life" Documentation

The best option is still the measured mile method of productivity analysis. Projects that started and gained progress before COVID started may provide a usable basis for an unimpacted measurement of productivity. This unimpacted productivity measurement could be compared to the impacted productivity incurred during COVID. The difference in productivity could be quantified and priced both prospectively and retrospectively. We are not going to restate the importance and implementation method of the measured mile, as that is covered in prior sections of this book. We simply want to emphasize its importance in the COVID world.

There may also be instances where projects were bid on, accepted, and started prior to COVID, but there is not a measurable comparison of work due to the time-related proximity of the project start and COVID. In these circumstances, another method may be more suitable.

Given the pervasive nature of COVID effects, the documentation of the specific effects may be a more cumbersome process. We encourage project teams to frequently create "day in the life" scenarios that document what downtime, work stoppages, productivity losses, and other COVID impacts are incurred on a daily basis. An example of a "day in the life" timeline for an electrical journeyman may look like the following:

- 5:30 – Arrived at site parking area – on the clock
- 5:35 – Got in line to board buses. Only 5 allowed on each bus due to COVID protocols. Would normally be a 5-minute ride to the site drop-off point.
- 6:00 – Board bus after temperature screening and COVID questionnaire (20 minutes lost).
- 6:05 – Arrive at site drop-off point.
- 6:10 – Arrive at man-lift/vertical transportation. Wait in line because only three people allowed on lift at once (including operator). Normally, 15 people can ride lift at once, resulting in a 5-minute wait normally.
- 6:30 – Get on lift for transport to 9th Floor (15 minutes lost).
- 6:35 – Arrive at 9th Floor. Fifteen-minute briefing on COVID requirements, spacing requirements, more temperature testing, and visual observation for signs of sickness (15 minutes lost).
- 6:50 – Daily work; break out and prep tools.
- 7:00 – Start productive work.
- 9:15 – Meeting with foreman; worker on 9th Floor with mechanical subcontractor tested positive yesterday. We are required to pack up hand tools, take with us, and go get rapid test.
- 12:30 – Return to 9th Floor and resume work. Rapid tests for 4 crew members were negative (3.25 hours away for testing minus 0.5 hours for lunch = 2.75 hours lost).
- 15:00 – Stop work for day; gather hand tools for transporting to vehicle.

- 15:10 – Arrive at lift elevator and wait due to COVID limitations (normally back on ground in 5 minutes).
- 15:45 – Lift arrives at ground (30 minutes lost); waiting for bus with COVID limitations (normally 5 minutes from bus to parking area at this time).
- 16:10 – Arrive at parking area – off the clock (20 minutes lost).

This preceding example identified a total of 10 hours and 40 minutes from the moment the electrical journeymen arrived at the site to the moment they left the site, equating to a total of 640 minutes at the site. The journeymen documented each "lost time" incident related to COVID, resulting in total COVID-related lost time of 265 minutes (41.4%) on that day. This information can be provided to the project management staff for tracking and, along with the company safety representatives tracking the COVID testing of staff, the company could quantify the number of lost hours due to COVID protocols and testing requirements.

The preceding is an example of a "day in the life" scenario that can be documented by the workers. It would appear abnormal to expect a shutdown of 2.75 hours each day for the crew. This would be an anomaly that the team should account for, but the remainder of the time impacts are applicable to other electrical staff at the site. Multiplied by the number of staff at the site, the analyst could quantify the total hours of nonproductive lost time per day due to COVID, then simply multiply by the labor rates and appropriate multipliers to obtain the cost of that lost time.

Of course, as the COVID requirements change, the documentation will require updating. The electrical superintendent can arrange meetings with project management and determine when requirements change and, as a result, a new documentation of "day in the life" is required.

The "day in the life" scenario is not an alternative to the measured mile. Rather, it can be used as a supplement to explain the causal nature of the various impacts, and why so much "lost time" is happening for the craft labor. The preceding example demonstrated that out of a total of 640 minutes, 265 minutes were nonproductive due to COVID on that day, resulting in COVID losses of 41.4% of all minutes expended for that crew. The more detailed information contractors have, the better they can track and, if possible, adjust and attempt to reduce productivity losses.

Comparisons to Similar Projects

For projects with no measurable "unimpacted period" pre-COVID, the best alternative to the measured mile may be a comparison to a similar project. The comparison to a similar project is different from a measured mile because there will be differences between the subject projects used to measure unimpacted and impacted productivities. The analyst should seek to minimize these differences and use the project most-similar to the COVID project, identify the differences and, when necessary, make reasonable adjustments for those differences. Again, our recommendation would be to use this method in circumstances when there is not a suitable time period available to measure the unimpacted period on the subject project.

Other Measurement Considerations

Given the nature of the COVID impacts and the level of variability between different projects and requirements, we currently do not believe there are applicable studies that could provide an accurate measure of lost productivity amounts due to COVID. We realize that the ability to track and monitor large amounts of data is being greatly improved and, at some time, there may be an applicable study in the future. The studies observed to date have been too generalized and relied too much on subjectivity and were qualitative in nature. Furthermore, you may understand our reservations with any industry study on COVID, given the track record of prior studies that were allegedly acceptable to the industry. As such, our recommendations for proactively accounting for productivity losses are the same as in other areas of this book. COVID simply took the importance of tracking productivity in real-time and added greater importance because of the pervasiveness of the effects. In short, the investment in process and tracking is time well spent.

The Way Forward

As of this publication, American owners and contractors have survived the initial onslaught of COVID, then Delta and now Omicron. We do not know what will come next, but our past experiences demand that our next set of contracts account for the possibility of that which we recently experienced. Owners and contractors alike should consider the extent to which labor and supply chain scheduling problems appear to have dissipated or are sustained.

Create models based on assumptions regarding such problems and identify performance parameters in the contract. Do this with the understanding that the owner is paying for that level of risk identified in the parameters set forth in the contract. For example, the contract will have assumptions that inflation will increase X% over each year of the contract. The contractor will bear the risk for anything over those parameters. Some contractors might want to push X% to an unacceptably high number during the contract negotiations, which may be unnecessary. Instead, if the parties agree that if inflation exceeds Y%, the parties could split the costs of inflation equally.

Once the contract has been negotiated, it behooves both parties to do that which we have been recommending throughout this book: (a) monitor the scope, schedule, and budget; (b) measure the work performed with the measured mile approach; and (c) mitigate mistakes, mismanagement, and monetary losses as effectively and efficiently as possible.

10

How Construction Disputes are Resolved

Owners, contractors, and subcontractors do not start the construction process with a desire to win a subsequent lawsuit. Instead, all the parties want to get the project built on time and on budget, and owners certainly want to avoid any shortcuts that could adversely affect the life and use of the constructed facilities. Despite those initial good intentions, many projects end up in one of two forms of dispute resolution: litigation or arbitration.

To succeed in court, to win an arbitration, or to successfully settle a construction dispute usually requires the selection and effective use of qualified expert witnesses. Expert witnesses are crucial to a party's success because such witnesses are deemed to be objective and very knowledgeable about the areas of their testimony. However, not every witness will be certified as an expert by the trier of fact. The United States Supreme Court, in a series of opinions starting with the 1993 decision in *Daubert v. Merrill Dow Pharmaceuticals,* 509 U.S. 579 (1993) and continuing through *Kumho Tire v. Carmichael,* 526 U.S. 137 (1999) six years later, articulated much more specific requirements that a witness must meet before that witness will be certified as an expert. The requirements for a witness to be certified as an expert, particularly in a construction dispute are the subject of Chapter 11.

To understand the importance of qualified experts under *Daubert* and its progeny, it is first necessary to understand the litigation and arbitration processes.

Litigation is that situation where one party (the plaintiff) files a complaint against the other party (the defendant) in federal, state, or administrative agency courts (such as the Federal Board of Contract Appeals). The other side (the defendant) eventually files its answer. At that point, the Rules of Civil Procedure, the Rules of Evidence, as well as that court's local rules, dictate the actions of the parties. In many litigation matters, the parties spend a great deal of time and money obtaining and reviewing the records of their opponents. Thereafter, each side asks a lengthy number of

Ted Trauner, Chris Kay and Brian Furniss. Productivity in Construction Projects, (159–164) © 2022 Scrivener Publishing LLC

oral questions and answers (under oath) of those people that will potentially appear at trial, known as depositions. The process of acquiring documents and taking depositions is known as the discovery process. Included in that discovery process will be the depositions of the opposing party's expert witnesses. The purpose behind the discovery process is to provide each side with a good idea of what the opposing party's evidence will be at trial. The theory is that once the parties know what their opponents will be offering as evidence at trial, then the parties are in a much better position to settle their dispute instead of going to trial. Despite the purported benefit of the discovery process, that same discovery in construction cases can take years to complete and can cost the parties hundreds of thousands of dollars or more.

Precisely because litigation can take years to complete and is very costly, many parties agree to employ the arbitration process, to the exclusion of litigation, to resolve post-construction disputes. That agreement to utilize arbitration is included in the contract between the parties, prior to the commencement of construction, and prior to any dispute between those parties. Many construction contracts also require that the arbitration is conducted by the American Arbitration Association, which typically provides a list from which three arbitrators are selected by the parties. In other contracts, a process is set forth that permits each side to select one arbitrator, and those two people then select the third arbitrator. Some contracts may even specify the arbitrator or arbitrators who will be used with no input or selection by the other party to the contract. Three arbitrators is not a magic number, as some contracts may even specify a single named arbitrator to be used.

Precisely because parties agree to the utilization of the arbitration process while negotiating the construction contract, parties with leverage—such as owners with general contractors and general contractors with subcontractors—can even be more specific in dictating the terms of the arbitration. The application of a given State's law, and the site of the arbitration dispute, are commonly included in the contract clause providing the parties with the right to arbitrate.

Once one of the parties demands an arbitration, the arbitrators are selected pursuant to the terms of the construction contract. Those arbitrators will then determine the amount of discovery, if any, the parties can conduct. Generally speaking, there is less discovery in arbitration than in litigation, including fewer depositions. However, since attorneys are frequently selected as the arbitrators, and those attorneys are familiar with the benefits arising from the discovery process, it is possible that the arbitration discovery process in some cases could become as extensive (and expensive) as the litigation process. No matter the extent of discovery permitted by the arbitrators for witnesses with knowledge of the facts, the witnesses identified by each party as experts are usually deposed prior to the commencement of the arbitration.

There are several similarities between trials and arbitrations. First, once the complaint has been filed to initiate the litigation process, the parties are usually required to meet and agree on the timetable for the completion of discovery in the case. This agreement is thereafter submitted to the court and a pre-trial order is executed by the judge. Later, usually after the time allotted for discovery in the pre-trial order has passed, the case is set on a trial docket before the judge assigned to the case—and the parties wait for their case to be heard in a courtroom by the trier of fact (either a judge or jury). The trial judge controls the scheduling of the trial. In an arbitration, a similar scheduling process is followed. Once the arbitrator(s) have been identified, they meet with the parties' attorneys and agree on a more limited discovery phase, as well as a date for the arbitration hearing. That date (or dates) is usually dictated by the availability of the arbitrators.

Second, when the trial or arbitration starts, the parties present their evidence, including the facts of the dispute and expert testimony. As the authors have emphasized throughout this book, the facts are the most important element in obtaining a successful result, either before a dispute goes to litigation or arbitration, or once the parties enter a structured dispute resolution procedure. Those facts also serve as the foundation for the expert's opinions. But the expert is very important, as parties would not be in this dispute if one side had all of the facts in its favor and the other side had no such facts. The expert helps the trier of fact understand which aspects of the competing set of facts are most important, and how those particular set of facts also justify a determination of what amounts, if any, should be awarded to the parties. The rules regarding the admission of evidence, including the testimony of experts, is dictated by the law of the State that the parties agreed to follow in their construction contract.

Third, and most important, the decision-making power does not reside with the parties. Instead, it resides with a judge, a jury, or a panel of arbitrators. But there is a way for the parties themselves to control the outcome.

In litigation, the court's local rules typically call for the two sides to engage in mediation after most, if not all, of the discovery has been completed. Mediation is ordered by the court to see if a settlement can be achieved without a trial. The parties choose a mediator, typically a former practicing attorney or judge. In mediation, each party attends with his/her attorney. While in the same room, both parties listen to the legal and factual arguments made by opposing counsel. Then the mediator separates the parties into different rooms and engages in "shuttle diplomacy" until a settlement is reached, or the parties declare an impasse.

Most cases are settled during the mediation process for several reasons. First, each side feels as if it has been given a chance to "have their day in court," as their claims have been presented to a neutral third party. Second, the parties have been forced to listen to the weaknesses of their case, first by the opposing party's attorney and then later by the mediator as part of the "shuttle diplomacy." Third, the parties must voluntarily agree to the settlement terms negotiated in the mediation process. If the parties do not both agree to the terms, then the mediator declares an impasse. In contrast to a negotiated settlement created and agreed to by the parties in the mediation, those same parties lose that decision-making control when the case goes to trial. The parties are reminded by the mediator that if a settlement is not reached, then the ultimate decision will either be in the hands of a jury that was specifically chosen because they know nothing about the topic, or in the hands of a judge that may be disappointed the parties did not achieve a resolution in the mediation.

Mediations have become so successful at resolving construction litigation that many arbitrators now require or encourage mediation as part of the arbitration process. Over the years, the authors have witnessed a growing number of contracts that require the parties to employ mediation prior to an arbitration hearing. Some of the parties to those contracts are going even further to control the parameters of the dispute resolution process. Having been unpleasantly surprised by the expansive amount of discovery permitted by some arbitrators or having been pleased with the efficient and equitable way a mediator was able to resolve prior disputes, parties are now beginning to identify specific people—in their construction contracts—to serve as mediators.

As a result, if a party seeks to "win" its litigation case or arbitration, that party will likely be "winning" a negotiated settlement—typically achieved in a mediation—as the vast majority of construction cases are settled at mediation or shortly thereafter. We do not see that well-established trend changing anytime in the COVID and post-COVID world.

As a result of this trend, the key to a party's success is the extent to which the case is settled on their terms. While acknowledging that the facts are always important, the best way to achieve the most favorable settlement is for that party to prepare to win their case even before a formal claim is filed. A party should start the process while the construction project is still ongoing. In particular, each party should (a) require its team to keep timely, comprehensive and well-documented records; (b) hire an experienced construction attorney when it appears that a significant dispute is brewing and either needs to be resolved quickly or by means of subsequent litigation or arbitration; and (c) ensure that the expert witness selected by the attorney is well versed in the facts of the construction project and meets the *Daubert* standards.

The team: Every party in the construction process should insist that its team is keeping timely and comprehensive records. It is the best business approach, whether or not a claim is subsequently filed. As set forth in these chapters, the best way to measure productivity, and act in a timely manner if inefficiencies occur on the job site, is to keep such records and review them on a periodic basis. Such a practice is the best way to manage the project, the best way to make timely changes if necessary, and the best way to prepare a case for litigation or arbitration. With the prudent use of smartphones, a party can make daily entries in software programs that measure progress; harkening to the old adage that a picture is worth a thousand words, those same smartphones easily facilitate the generation of daily or periodic photos that demonstrate the progress, or lack thereof, at the jobsite.

The attorney: Every party should also consider engaging an experienced construction attorney whenever meaningful scope, schedule and/or budget issues arise during the construction. Don't wait until the problem has grown to such a size that litigation or arbitration is inevitable. We also caution parties not to think they can solve the problem by themselves, and thus avoid paying an attorney for his/her fees. Instead, we recommend a proactive approach, with the party viewing the attorney as an asset that helps solve the problem before it gets any larger. This approach requires the attorney to understand shortly after the outset of his/her retention what are the critical issues, and the documents that support their client's claim.

Some parties may initially blanche at the idea of adding the cost of an attorney at this juncture, but this approach can actually be financially beneficial. Since most construction disputes end in settlement, the party's attorney can collect the relevant and persuasive documents as the construction progresses. Under this approach, that party will be much better prepared to sit down with the opposing party (and his/her attorney), thereby increasing the possibility that the dispute can be settled without initiating substantially costlier litigation or arbitration. Even if those efforts are initially unsuccessful and a claim is filed, a party's attorney (having previously been educated on the facts of the claim) will be in a better position to streamline the discovery process and push for an earlier date for trial or arbitration hearing, and thus potentially obtain an earlier date for a mediation and resolution of the matter. In every step of this process, an experienced and well-informed attorney can save money for a proactive party.

The expert: Every party should insist that its attorney retain an expert that will meet the current criteria for testimony at trial. Due to the fact that the criteria have changed over the last three decades, and the application of that criteria to experts in construction cases has become far more complex, the following chapter is devoted to this topic.

11

The Selection and Use of the Expert Witness

The Selection of an Expert

At some point in the process, each party needs to determine whether they should retain an expert witness to analyze their position in the construction dispute. For the reasons set forth in the preceding chapter, it is often prudent to retain an expert while the project is ongoing, even before a claim is actually made. At the other end of the spectrum, the retention of an expert must be completed before the end of discovery, as the expert needs to offer his/her report, and likely testimony, before the trial or arbitration is conducted. In many instances, the expert witness is the last person to be deposed by each side during discovery. It is not a coincidence that most cases settle only after the deposition of the experts.

Why should an expert be retained? Especially in construction cases where there is a claim of labor utilization inefficiencies or lack of productivity, the parties cannot rely solely on their payroll records and the testimony of their personnel. Precisely because there are so many variables and competing factors that occur at the job site—which may or may not adversely affect the productivity of the workforce—an expert witness is usually needed to help the trier of fact understand what happened, explain how that disputed event or changes led to a lack of productivity (or not), and reasonably quantify the estimated resulting loss of labor productivity and damages (if any).

An experienced and objective expert witness can offer the most comprehensive, persuasive and even dispositive testimony, the kind of testimony that can have a significant impact in the resolution of the case—either at mediation, trial, or arbitration. To be persuasive, the expert must quickly become fully knowledgeable about the facts and issues of the case and perform the most acceptable, recognized analysis or method to reach his/her

Ted Trauner, Chris Kay and Brian Furniss. Productivity in Construction Projects, (165–190) © 2022 Scrivener Publishing LLC

conclusions. A party's attorney and expert should work collaboratively on all aspects of the case, for it is the expert—not the attorney—that will offer the expert opinions to the trier of fact. For this reason, the attorney—rather than the party—typically selects the expert witness.

What kind of person should be selected as an expert? Retaining an expert is much more than just selecting someone with university or industry "credentials." The attorney needs to retain an expert that will be:

- Qualified as an expert under the *Daubert* standard and the applicable rule of the Federal Rules of Evidence (Rule 702), thus permitting the expert to testify at trial.
- Knowledgeable about the type of construction issues and the facts of the particular dispute; and
- Persuasive to the trier of fact by providing opinions that are based on the actual facts of the construction project in dispute and applying those facts to reliable construction principles.

The Criteria an Expert Must Meet: The *Daubert* Standard

Rule 702 of the Federal Rules of Evidence initially created in 1975 stated, in part:

> *"If scientific, technical, or other specialized knowledge will assist the trier of fact to understand the evidence or to determine a fact in issue, a witness qualified as an expert by knowledge, skill, experience, training, or education, may testify thereto in the form of an opinion or otherwise..."*

In the 1993 United States Supreme Court case of *Daubert v. Merrill Dow Pharmaceuticals*, 509 U.S. 579 (1883), the parents of children born with birth defects sued the drug manufacturer, claiming the drug Bendectin had caused the birth defects. There was a battle between the *Merrill's* expert, who argued that there were no scientific studies tying the birth defects to the drug, and the parents' expert, who argued a novel theory supported by data from animal studies. Under a prior United States Supreme Court ruling, the exclusive test for the admissibility of expert testimony was whether the opinion expressed by the witness was "generally accepted" in the field.

In *Daubert*, the U.S. Supreme Court ruled that "general acceptance" was not the standard under Rule 702 of the Federal Rules of Evidence. Instead, the trial court must ensure that an expert's testimony rests on a reliable foundation of scientific knowledge before it is to be considered by the trier of fact. The critical question for the court is whether the witness' underlying reasoning or method is scientifically valid and properly applied to the facts of this case—not whether it was generally accepted in the industry.

Prior to this case, it was typical for the trial courts to simply certify each party's experts after learning of their credentials and determining that the methods by which the evidence was obtained was generally accepted by experts in that particular field. Then let the jury decide. The Court's certification of a witness as an expert was so routine that experienced trial lawyers would sometimes stipulate to the opposing party's expert, such that the jury would not learn of that witness' academic or scientific bona fides.

But all of that changed with the new standards for admissibility of expert testimony, first expressed in the 1993 *Daubert* decision and in two cases later that same decade, *General Electric v. Joiner*, 522 U.S. 136 (1997) and *Kumho Tire v. Carmichael*, 526 U.S. 137 (1999). This trilogy of cases forms what is now commonly called the *Daubert* standard.

After *Daubert*, trial courts have continued to take a much more active role as gatekeeper, rigorously evaluating whether the expert would be permitted to offer his/her opinions to the jury. Since this Supreme Court decision, it is now more common for trial courts to conduct a *Daubert* hearing prior to trial, where the judge determines the extent to which a party's expert can offer his/her opinions to the trier of fact.

In applying this ruling to construction cases, a party's expert can no longer simply cite his/her credentials and expect to be certified as an expert and testify at trial. Instead, the expert's knowledge of the subject matter of the dispute, the actual facts that occurred, and the recognized method that serve as the underpinnings of the expert's opinions will likely be challenged by the opposing counsel and carefully examined by the judge.

In *General Electric*, the experts battled over whether *Joiner* had been exposed at his workplace to certain chemicals manufactured by *General Electric* that allegedly "promoted" his subsequent lung cancer. The trial court determined that the expert had failed to set forth a reliable link between *Joiner*'s exposure and his subsequent cancer, in that the expert's testimony was based on studies that were very dissimilar to the facts in the case. In addition, some of the studies upon which the expert relied specifically did not make the causation link between the exposure and the subsequent cancer, the very essence of the expert's opinion. As a result, the trial court excluded the testimony of the expert, thus effectively ending *Joiner*'s case.

The Supreme Court also stated that discretion should be given to the trial court judge when sitting as the gatekeeper, as the judge is responsible for determining whether an expert's opinions are not only relevant, but reliable. Here, the Supreme Court held that the trial court did not abuse its discretion in excluding the expert's testimony.

What is important about this decision for experts in construction cases is the Supreme Court's criticism of *Joiner*'s expert for relying upon studies that did not conclude the causal link between the alleged wrongful event (exposure to certain chemicals) and the resulting damage to the party (cancer). As a result, these studies were an insufficient basis for an expert's opinion. Similarly, a construction expert witness' sole reliance on industry standards, such as those of the Mechanical Contractors Association of America, can lead to similar debilitating problems.

In *Kumho Tire*, the Supreme Court held that the *Daubert* standard applies to all expert testimony, including testimony that is non-scientific. This case involved the opinions of a tire engineer as to why a tire exploded while in use, thus causing injury and death to those in the vehicle. The tire manufacturer argued that *Carmichael*'s expert should be excluded because he wasn't a scientist. The fact that the testimony was not "scientific" did not, alone, invalidate the expert's opinion, the Supreme Court held, because Rule 702 also applied to "technical and other specified knowledge." Thus, engineers and others with considerable construction experience can be deemed an expert under Rule 702.

Equally important to construction disputes is the Supreme Court's focus on whether the expert's methods reliably determined what caused the tire to explode. Based on his experience and inspection of the tire, the expert testified that there was no other possible explanation for the tire failure. He did not explain the cause of the tire failure, but only that he could not come up with any other explanation. That testimony "in the negative" failed the *Daubert* test, for it was not based on facts that would support the opinion. Instead, the Court found the expert's opinion to be nothing more than an unsupported, conclusory statement. As a result, the testimony was excluded. To the extent that a construction expert's testimony is based on his/her opinion that is not supported by facts, it will now face a similar fate. There must be a clear and reliable causal connection between the alleged wrong (such as a design error or a change order) and the alleged resulting damage (such as labor inefficiencies), and that connection must be proven by reliable means.

These three U.S. Supreme Court cases have had an impact on the admissibility of expert testimony in construction cases. In particular, the past practice of a construction expert's utilization of "industry standards" has been significantly curtailed. A party's expert takes great and unnecessary risks if he/she relies heavily upon "industry standards" rather than basing their testimony primarily upon their analysis of the facts of the case. Since the application or the *Daubert* trilogy of cases regarding the admissibility of expert testimony, courts have frequently and more consistently rejected a purported expert's testimony when it is solely or substantially based on the application of "industry standards."

As we have written throughout this book, the parties and their attorneys need to have a full and commanding understanding of the facts of the construction process. There is simply no substitute for hard, detailed, and comprehensive work. With the widespread implementation of the *Daubert* standard, that same work ethic is now required of the expert witnesses as well. A construction expert cannot simply rely upon some factually limited, general, or faulty study, refer to it as an "industry standard," and then hope that the testimony will prove successful in court or in an arbitration.

Since those three United States Supreme Court decisions, the *Daubert* standard has been applied in all federal courts. Federal Rule 702 has been twice amended to provide greater clarity to the parties, the witnesses, and the judges. Rule 702 now states:

> *A witness who is qualified as an expert by knowledge, skill, experience, training, or education may testify in the form of an opinion or otherwise if:*
>
> (a) The expert's scientific, technical, or other specialized knowledge will help the trier of fact to understand the evidence or to determine a fact in issue;
> (b) The testimony is based on sufficient facts or data;
> (c) The testimony is the product of reliable principles and methods; and
> (d) The expert has reliably applied the principles and methods to the facts of the case.

In addition, the vast majority of states have applied these same standards, or modified versions, in their jurisdictions. But there are still a handful of states that have not adopted the *Daubert* standard. As a result, the parties, their attorneys, and expert witnesses should learn and thereafter follow the applicable legal criteria for admissible testimony in the states in which they are working.

The Application of the *Daubert* Standard in Construction Cases: The Benefits of the Measured Mile Method

Applying the *Daubert* standard and Rule 702 to construction disputes, one can see why experts utilize the measured mile approach whenever possible, for it is the most likely path to having their testimony presented to the trier of fact. The measured mile approach is predicated on the actual facts of the particular construction project, where the expert compares the productivity in certain parts of the job site before there was a change of events to the productivity of other parts of the same job site adversely impacted after those events occurred. Alternatively, the analysts may compare the facts of similar projects, one where there was an intervening change order or event, and the other where no such disruption took place.

In both situations, a qualified expert has the opportunity to isolate the specific cause of a lack of subsequent productivity and then reasonably quantify its cascading impact of labor inefficiency on the job's scope, schedule, and budget.

Many courts and administrative boards have found the measured mile to be the most reliable method to quantify labor inefficiencies, while simultaneously acknowledging that there cannot be an exact calculation of labor inefficiencies. For example, in *Clark Concrete Contractors, Inc. v. General Services Administration*, G.S.B.C.A. No. 14340, 99-1 B.C.A. (CCH) 30280, 99 WL. 143977 (Gen. Services Admin. B. C. A. 1999) the Board of Contract Appeals found that the measured mile was a reliable means of determining the cost of labor inefficiencies created by the owner, even if the amount could not be precisely calculated. The case involved the construction of the Washington D.C. field office building of the Federal Bureau of Investigation. After construction had commenced, there was a bombing at a federal building in Oklahoma City, thus prompting the government's architects to make "massive blast design changes" to make the FBI building more capable of withstanding a similar bomb blast.

The contractor sued for the additional costs and delays associated with the design changes. After making a determination that the GSA was responsible for 185 days of delay, the Board considered the testimony of the parties' experts.

The contractor's expert utilized the measured mile method. He compared the productivity achieved in the unimpacted period with that achieved in each of the impacted periods. Specifically, the expert determined that with respect to the forming of floor slabs, there was a productivity rate of .042 man-hours per square foot during the unimpacted periods, and a .048 rate per square foot during the impacted periods. The expert subtracted the achieved productivity rate during the unimpacted period of time from the actual productivity rate during the impacted period of time to measure the inefficiency. He then multiplied that figure by the quantity of slabs formed during each period to determine the number of lost man-hours. He multiplied that number of lost man-hours by the average hourly rate for members of the forming crews, to calculate the total labor inefficiency cost for forming slabs. The expert conducted the same kind of analysis for the work associated with the stripping of the floor slabs, the forming and stripping of columns, and the forming and stripping of stairs. The expert concluded there was $1,181,508 in inefficiency labor costs.

The government's expert contended that the contractor's measured mile analysis was not valid, as the concrete work performed after the design changes was not identical to the work performed before the design changes. Instead, the government's expert contended the contractor's calculations were nothing more than a modified total cost approach, and inherently unreliable.

There is no dispute that the work was not identical. The very essence of the design changes was to make the building better able to withstand a bomb blast—through the use of more concrete. But that fact did not negate the ability to reliably employ the measured mile approach. The Board's analysis is instructive in several respects, as it found:

> *"The purpose of a measured mile analysis is to permit a comparison of the labor costs of performing work during different periods of time, so as to show the extent to which costs increased from a standard during periods impacted by certain actions. See U.S. Industries, Inc. v. Blake Construction Co. 671 F.2d 539, 547 (D.C. Cir. 1982); Stroh Corp. v. General Services Administration, GSBCA 11029, 96-1 BCA 28,165, at 141,132. GSA is correct in asserting that the work performed during the periods compared by [the contractor] was not identical in each period. We would be surprised to learn that work performed in periods being*

compared is ever identical on a construction project, however. And it need not be; the ascertainment of damages for labor inefficiency is not susceptible to absolute exactness (emphasis added). See Electronic & Missile Facilities, Inc. v. United States, 416 F.2d 1345, 1358 (Ct. Cl. 1969). We will accept a comparison if it is between kinds of work which are reasonably alike, such that the approximations it involves will be meaningful. We conclude, as explained following, that [contractor] has generally used a measured mile analysis consistent with this objective, but that some of the adjustments made in the course of the analysis are not justified."

As an additional argument, the government's expert conducted his own version of a measured mile analysis. The Board compared the measured mile analysis of the two competing experts, finding computation errors in the contractor's measured mile analysis, but significant differences in the allegedly "comparable" work set forth in the government's measured mile approach. The Board wrote that "although no perfect comparison involving different floors during different periods of time could be devised for the project," (it) found the measured mile proposed by the contractor to be the most reliable one. Having so ruled, the Board nonetheless made some adjustments to the calculations performed by the contractor's expert, which reduced the claim from $1,181,508 to $1,082,490.

There are four important lessons to be learned from this decision, those being:

1. The Board permitted the comparison of work even though it was not identical. It even opined that it doubted such comparable work could ever be identical.
2. The Board found that the claim for labor inefficiency need not be absolutely exact; rather it is an "approximation."
3. The court will adopt an expert's measured mile method where the work compared in the "before and after" time periods is sufficiently similar for reliable comparative purposes. Here, the contractor's expert's comparison of arguably comparable work was much more logical, and much more tied to the actual facts of the project, than the work compared by the government's expert—even though both experts claimed they were correctly employing the measured mile method.

4. Finally, when a court or board conducts such review of a successful party's claim based on the measured mile, it can and will make adjustments to these calculations, thus reducing—but not negating—the amount a party may recover.

As reflected in *Clark Concrete* and the cases cited therein, the measured mile is the preferred method to identify and calculate labor inefficiencies, when performed in a reliable fashion. Whether a measured mile analysis is deemed reliable is often determined by the degree to which the expert has compared two reasonably similar periods of time, work, and accurately measured and performed a productivity analysis.

As a result, parties in construction disputes frequently retain experts to argue the measured mile approach. However, when those experts are retained after the completion of the project, the expert is often faced with a lack of data to support such an analysis. The authors have recommended in prior chapters that it is wise to promptly retain experts when problems arise that potentially adversely affect productivity—primarily to solve or minimize the costs of such inefficiencies. This approach also provides the expert with the opportunity to direct the party to engage in additional or differentiated documentation (as the project proceeds) that can provide support for a measured mile analysis at the end of the project.

Precisely because courts and agency boards recognize the measured mile as the preferred approach, many experts attempt to base their testimony on this method, even if the facts of their case do not support such an analysis. Sitting as the gatekeeper, courts have rejected expert opinions based on the measured mile for a variety of reasons, including but not limited to:

(a) Inadequate or flawed labor records;
(b) An expert's utilization of non-comparable work;
(c) The use of a sample size of work that is comparable but demonstrably too small; or
(d) Because the expert's credibility is in doubt due to the subjective selection of the periods of productive and nonproductive work compared in the report.

Some of the grounds for excluding such testimony may be beyond the control of the parties. But there are things a party can control. The party can quickly retain an expert, address the intervening change and its associated problems, and carefully consider whatever changes the expert recommends regarding the quality, specificity, and quantity of a party's records.

Those steps that a party can control may make a significant difference in getting the judge to permit the expert to testify using the measured mile method, which is yet another important reason to retain an expert as soon as possible in the construction process.

The Application of the *Daubert* Standard in Construction Cases: The Pitfalls of the "Industry Standards" Methods

Suppose there is no way the facts lend themselves for an expert to successfully present the measured mile method? In many cases, experts have turned to "industry standards" as an alternative. Industry standards include the publications of the Mechanical Contractors Association of America (MCAA), the National Electrical Contractors Association (NECA), the Business Round Table (BRT), and the U.S. Army Corps of Engineers (ACOE) among others. To the extent such publications served as the basis for expert testimony prior to the creation of the *Daubert* standard, such testimony is now more rigorously scrutinized by courts and often excluded in construction cases.

The 2019 federal district court case of *Trane U.S. Inc v. Yearout Service LLC* No. 5-17-CV-42-MTT, 2019 WL 2553100 (M.D. Ga. June 20, 2019) is the most recent published opinion regarding the admissibility of expert testimony in construction cases. This federal district court decision illustrates the problems that a party will likely face when expert's testimony is solely or substantially based on industry standards.

The dispute involved the general contractor for a renovation at a U.S. Air Force Base and *Yearout*, a subcontractor responsible for providing a "turnkey mechanical and plumbing system." *Yearout* sought over $280,000 in overtime costs due to alleged inefficiencies created by the general contractor. *Yearout*'s expert's opinions were based on the estimated overtime inefficiency costs. In support of his opinion, the expert used payroll records and a "productivity index table," as well as guidance from MCAA Management Methods Bulletins OT1 and OT2-2011.

The trial court conducted a *Daubert* hearing to determine if *Yearout*'s expert witness testimony should be permitted at trial. After hearing that testimony, and carefully reviewing the MCAA standards and its underlying support, the judge concluded the expert's testimony was "seriously flawed." As a result, the trial court granted the opposing party's motion to exclude the testimony of *Yearout*'s expert.

The trial court first addressed the MCAA overtime inefficiency standards, observing that "the overriding purpose of the MCAA bulletins is to provide mechanical contractors with negotiating tools: primarily prospectively, to obtain more favorable terms, but also retrospectively to support claims for additional compensation. The bulletins do not, nor do they purport to be, peer-reviewed studies determining whether overtime and shift work 'inefficiencies' exist and, if they do, how these 'costs' can be calculated." Before addressing the foundational basis of the MCAA bulletins, the judge noted that the industry has a more precise standard to calculate such costs, that being the measured mile approach.

The trial court then evaluated the MCAA bulletin and its underlying support, which comes from other industry standards—namely the NECA, BRT and ACOE studies:

> *"The problems with [Yearout's expert's] testimony start with the MCAA bulletin, which bases its percentages table on the four studies. The bulletin, however, provides almost no information about the studies themselves. The first study, the Business Round Table, examined data from twelve weeks of one construction project to "demonstrate[ ] that, in general, inefficiency increases as the overtime schedule extends in duration." Doc. 158-4 at 129. A second source of the MCAA numbers is a study by an electrical contractors' trade group, NECA, based on a survey of electrical contractors. Id. The MCAA argues that the NECA study translates to mechanical contracting because "[i]t is a generally accepted axiom in the construction industry that efficiency impacts sustained by the mechanical trades are similar in nature to the inefficiency impacts sustained by the electrical trades given reasonably comparative adverse conditions." Id. n.9. Nowhere is this "axiom" supported. Third study, authored by an engineer, appears to only have data for overtime loss for 50-hour weeks. Id. at 129-133. "Appears" because the Court, short of purchasing a copy of the study, has no way to tell what the study actually did.3 The fourth study, by the Army Corps of Engineers, "was widely used ... until the Corps formally withdrew this publication several years ago for unspecified reasons." Id. at 130. Despite the ACOE's formal withdrawal of their study, the MCAA finds it "noteworthy that Publication EP 415-1-3, which contained the Corps' overtime study, has never been repudiated by the Army Corps of Engineers." Id. (emphasis added). The distinction between "withdrawn" and "repudiated" eludes the Court.4*
>
> The MCAA OT1 methodology, therefore, depends on four studies: one which only focused on one project, another based on a survey

of electricians, and another of which has been withdrawn by its publisher. And none of which has been made available to the Court."

Having exposed the weaknesses of the industry standards, the trial court then focused on the expert's work, or lack of work, related to these studies:

"Without access to the underlying studies, the Court had to rely upon [the expert's] knowledge of the studies. However, [the expert] has not even read the studies.5 So even if the MCAA's asserted "axiom" that the relationship between overtime and inefficiency is the same for electrical contractors and mechanical contractors were true, and even if the Court were to accept the MCAA's contention that the COE study has only been withdrawn, not repudiated, the Court still could not conclude that this methodology has any reliability."

Equally significant, the trial court observed that the expert's opinions were not based on his personal investigation of the project. Instead, the expert's testimony was "based on an inconsistent and one-dimensional application of a trade group's statistical table based on a small set of limited or withdrawn studies that have not been made available to the Court or even read by [the expert]. And *Yearout*'s brief defended the testimony by citing cases which used a different methodology altogether. The testimony is clearly inadmissible under the standards of Rule 702 and *Daubert*."

The *Trane* decision is not the only case where an expert's reliance on industry standards has been excluded because it lacked sufficient foundation. In fact, most of the reported decisions since the *Daubert* standard was completed with the 1999 Supreme Court's decision in *Kumho Tire* have rejected expert testimony based on each of the "industry standards" rejected by the trial court in *Trane*. Those industry standards, and the courts' assessment of their reliability, are set forth in the following pages of this chapter.

But before addressing how courts and administrative agencies have accepted or rejected testimony based on "industry standards," it is important to explain the relevance of previously rendered cases. As noted above, the trial court stated that "*Yearout*'s brief defended the testimony by citing cases…."

What does it mean to "cite cases"? Each party's attorneys' will make arguments to the trier of fact that their client's arguments are based upon legal principles and/or similar factual arguments contained in previously published cases that demand a similar result in the matter in dispute. They "cite" those cases supportive of their theories of the case. By contrast, the opposing party's attorneys will argue that those cited cases by the other side are no longer "good law" because they have been overturned on appeal or the relevant statute or rule has been amended, the cases have not been adopted or used by other courts, and/or those cases are not applicable to the facts of the case in dispute. Instead, the opposing attorneys usually assert there are different cases that are much more relevant to the facts in dispute. One can assume that if the legal dispute goes on for several months or years, the attorneys on each side are arguing different case decisions in support of their client's position.

As a result, it is extremely important for the attorneys to conduct legal research to determine the extent to which courts and administrative agencies have subsequently relied upon, rejected, or ignored previously published decisions. The most fatal of flaws is to rely upon a lower court decision that has thereafter been overturned on appeal. Almost as unfortunate is to rely upon a case decision that is not cited or utilized by other courts when explaining the basis for its decision, or distinguished by those courts as to be rendered irrelevant. The entire basis for a case, as well as the credibility of the attorney, can be effectively destroyed if the opposing counsel demonstrates to the trier of fact that their opponent is relying upon case law that is no longer applicable. To avoid such blunders, attorneys have used the Shepard's Citations Service (also known as "Shepardizing" a case)—or otherwise conduct case decision "updates" on the internet—to determine whether a prior published decision has been affirmed or overruled on appeal, is cited by other courts in support of the ruling, distinguished by other courts that came to a different conclusion, or not even mentioned when similar facts are involved in a case before a different court or agency board.

It is equally important for the would-be expert to understand the adoption, rejection, or avoidance of a given decision by other courts and agency boards. As is true in many current contentious aspects of modern life, some judges can take unique or "activist" positions that may temporarily affect the law, but those positions may not thereafter be adopted by any other court or agency board. The authors have seen the same situations occur in construction cases, where a particular judge or agency board may render a decision utilizing an industry standard, only to see that decision be rejected or ignored by other courts. If an expert relies upon an individual decision that is subsequently overruled, distinguished, or ignored, the loss of their credibility will be as devastating as that of the attorney described previously.

Finally, it is important to understand when a case is decided. Some cases do not stand the test of time. For example, there were a few decisions that were rendered shortly after the *Daubert* trilogy of cases that may have given some weight to "industry standards," usually because the facts suggested it was fair to compensate a particular party. But when those decisions are not relied upon by any other courts or agency boards after the widespread implementation of the *Daubert* standards, the results in those earlier cases tend not to be dispositive or persuasive in subsequent cases.

At the opposite end of the time spectrum are those cases recently decided, such as the 2019 *Trane* case previously described. That decision reflected the breadth of current and well-established utilization of the applicable standards for assessing the admissibility of expert testimony—and the lack of judicial respect for "industry standards." As a result, large portions of that opinion can be used to similarly eliminate the testimony of any expert relying substantially on "industry standards." The court has literally provided a "blueprint" for the exclusion of that purported construction expert's testimony before the trial or arbitration.

Stated differently, everyone in the construction dispute process, including the parties themselves, should be wary of putting too much reliance upon one case decision that tends to support their arguments until that case has been Shepardized, or updated, to determine if any other court has found it persuasive. In addition, attorneys and parties should also be wary of a purported expert that primarily relies upon "industry standards." As demonstrated in the following sections devoted to each of the so-called "industry standards," the courts and administrative agencies have frequently (and with much greater regularity in the last two decades) chosen not to admit such testimony.

One final point about shepardizing or updating cases: some of the construction cases cited in this chapter have been favorably cited anywhere from 5 to 198 times. Only one case received a somewhat negative treatment, where the appellate court upheld in part and reversed in part. The change on appeal was based primarily on the manner in which the lower court had used one date from which to determine damages for six different changes that occurred over time, as well as how that court miscalculated home office overhead damages.

Industry Standards: Mechanical Contractors Association of America (MCAA)

Since 2000 there have been 16 reported decisions from U.S. federal courts (3 in total), state courts (2), and administrative agencies (11) regarding the admissibility of expert testimony that rely on the MCAA factors to prove labor inefficiencies. The courts in all three published federal court decisions have rejected expert testimony based on MCAA industry standards. Of the 11 published administrative agency board opinions, expert testimony based on MCAA has been rejected in six of them. In three cases, the testimony was accepted in part and rejected in part. In two other administrative cases, the testimony was admitted, but only because entitlement had already been admitted (*Appeal of Prince Construction, DCCAB No. D-1127, 2003 WL.21235618 (D.C.C.A.B. May 12, 2003)*) or where the expert had made his own adjustments to the MCAA factors for the specific mechanical work in dispute (*Hensel Phelps Company v. General Services Admin, G.S.B.C.A. No. 14744, 01-1B. C.A. (CCH) 31249, 2001 WL 43961 (Gen Services Admin. 2001), aff'd 36 Fed Appx. 649 (Fed Cir. 2002).*

Federal Trial Court Decisions Relating to MCAA

Turning first to the published federal decisions, in addition to the 2019 *Trane* decision (previously discussed) there is the 2015 decision in *North American Mechanical Inc. v. Walsh Construction Company and the 2005 case of Sunshine Construction & Engineering Inc. v. United States.*

North American Mechanical (132 F. Supp. 1064 E.D. Wisc.) was a dispute between the general contractor (Walsh) and an electrical subcontractor (NAMI) regarding the renovation and construction of new space at a hospital. The construction was phased in such a way to permit the hospital to continue to operate during the construction. Such construction phasing during occupancy can create scheduling challenges, and that was the case here—as scheduling problems plagued the project from its onset. At the end of the project, the electrical subcontractor claimed $1,747,326 in labor inefficiencies.

Pursuant to the terms of the contract, the dispute was presented to the court, rather than to a jury, for resolution. The parties agreed that the trial judge's Magistrate Judge would try the case. During the pendency of the case, Walsh moved for summary judgment. As part of that motion, Walsh also moved to bar the testimony of NAMI's expert witness. The court had doubts about the testimony, but since the case was being tried by a judge, the Magistrate Judge believed it best to permit the testimony and then give it the weight it deserved. (This approach is often used by arbitration panels as well as in judge-tried cases.) After considering all of the facts and testimony, the Magistrate Judge ruled in favor of Walsh. The federal district court rejected NAMI's subsequent appeal on several separate grounds. This case provides several important lessons.

1. The dispute was resolved in court, by a judge. The authors have noted the desirability of having language in the contract that specifies the use of selecting certain arbitration services or even individual arbitrators. We have found this approach minimizes time and legal expense associated with the resolution of the claim. Here, the parties contractually chose to

have the case adjudicated in a federal court, but by the judge rather than a jury. Such an approach can potentially create an advantage for certain parties. But it should be noted that this so-called advantage, if any, calls for some speculation on the part of the parties when the contract is drafted. Who knows as of the time the contract is signed whether a party's legal argument is more analytical in nature, and thus more likely to be understood more clearly by a judge, or which side would have a more equitable factual claim that could be embraced more readily by a sympathetic jury?

2. What does not require speculation is the second important lesson learned from this case: the terms of the contract will be applied to the parties. Here, the terms required NAMI to make specific claims within 7 days of discovering that such owner-requested changes would create additional costs for it. Although such a short time requirement seems severe in the abstract against this subcontractor, in this particular case the general contractor was required to bring such claims to the owner within 21 days, according to the terms of its contract with the owner. Thus, the subcontractor was on a short timeline to report costs for owner requested changes precisely because the general contractor was also on a short timeline of the owner for such requested changes. More important, the subcontractor contractually agreed to perform pursuant to those short timelines.

NAMI failed to follow that contractual term with specificity, and therefore the claim was denied. The authors cannot stress enough the importance of negotiating contractual terms with which both parties can successfully comply and perform. We have been advised by parties, once litigation has commenced, that they knew a disputed contractual clause like this one would be difficult to follow—but they agreed to such terms anyway because they feared they would otherwise lose the bid. In retrospect, some of these parties have acknowledged that their burning desire to get the work had effectively burned their legal rights in their subsequent dispute resolution proceeding. We recommend that if a party is going to agree to such terms in the written contract, the party should immediately hire all necessary personnel, and institute all necessary reporting requirements, to be able to fully comply with such problematic contract terms.

3. Third, it is critically important to keep contemporaneous and thorough records. Without such records, a party's expert cannot use the generally preferred approach of the measured mile to calculate labor inefficiencies. Instead, the expert is forced to use other methods with unacceptable, but not unexpected, results in court. In this case, NAMI's expert was forced to employ the total cost method precisely because NAMI did not have such supportive records. The total cost method computes damages as the

difference between the amount bid and the actual cost incurred. Courts have recognized that this approach can improperly reward a party that has either created an unrealistically low bid or has been inefficient in its own performance on the job. As a result, the total cost method has often been rejected, or at least viewed as the method of last resort.

In support of the expert's findings under the total cost method, the expert provided opinions based on the measured mile and NECA standards. Although the court had already concluded that the total cost method was without merit, it examined the other two methods to see if either one could justify a finding in favor of the subcontractor.

NAMI's expert sought to provide a measured mile expert analysis but admitted he did not have sufficient data to render an opinion that could stand on its own, and hence it was only offered in support of the total cost method. The trial court agreed with those admitted deficiencies, finding that only 30 man-hours of unimpacted labor was used for the expert's measured mile approach. But 30 man hours on a project that required 32,000 man-hours to complete [is] only 0.094% of the total work on the project—much too small a sample to be credible and persuasive.

4. Fourth, an expert's reliance on the MCAA factors—without specific job site factual support—is often a recipe for failure. Here, the subcontractor's expert relied on the MCAA factors in support of the total cost method, specifically the MCAA factor involving "beneficial occupancy" (where an owner, such as this hospital, operates the building during the construction process). The court noted that the expert incorrectly calculated such an impact on the entire project, even though some of the work was in an area that had not yet been occupied, that being the newly constructed areas of the expanded hospital. Beyond that, the court concluded that since NAMI knew that the hospital intended to continue its operation prior to entering into the contract with the general contractor, it should have factored such potential problems into its bid, and thus this MCAA factor was inapplicable to both the new and the renovated areas.

Perhaps more important was the court's observations about the MCAA factors when it wrote:

> *"Moreover, the MCAA bulletin states:*
>
> *These factors are intended to serve as a reference only. Individual cases could prove these to be too high or too low. The factors should be tested by your own use of them, since percentages of increased costs due to the factors listed are necessarily arbitrary and may vary from contractor to contractor, crew to crew and job to job."*

The court concluded NAMI's expert failed to take "the next crucial step of analyzing the specific conditions of the Project to assess the extent and impact of any condition to arrive at an appropriate inefficiency rate." In light of the nature of the expert's four offered opinions, his failure to analyze the specific job site conditions may well be attributable to the subcontractor's lack of contemporaneous and extensive data.

The *North American Mechanical* court cited *Sunshine Construction & Engineering v. U.S. 64 Fed Cl. 346 (2005)* with approval, a decision rendered in 2005 by the Federal Court of Claims, for the proposition that MCAA factors are not dispositive because they are general propositions. In *Sunshine*, the contractor's $915,872 claim against the Army Corps of Engineers involved the construction of an educational center and library at the MacDill Air Force base in Florida. While acknowledging that the architects' blueprints were substantially inadequate, the federal court rejected the contractor's claims. The case is noteworthy for three reasons.

First, it is important for a party's employees to testify as objectively as possible—especially since the party is using its witnesses to persuade the trier of fact. Sunshine's key fact witness was its Project Superintendent, who testified a substantial part of Sunshine's problems were due to his counterpart at the ACOE, a person that allegedly impeded progress and directed "quite a bit of hostility" towards Sunshine and its representatives. The court noted from the correspondence that the parties had several disagreements and communication issues. But the court heard and witnessed the testimony of that ACOE's Contracting Officer's Technical Representative, and found him to be a reasonable, if exacting, representative of the ACOE. The credibility of Sunshine's claims suffered because its Project Superintendent's personal frustration during the construction process carried over into the courtroom.

Second, as previously noted, construction disputes are often a battle of the experts. The court's analysis of the two experts underscores the importance of retaining the most qualified, thorough, and objective expert available:

> *"Key to defendant's successful proof against plaintiff's claims was the thorough, detailed testimony of Mr. Bolyard, a certified structural engineer who served as defendant's expert. Mr. Bolyard was also very knowledgeable in terms of the overall contract performance sequence. While Michael T. Midgette, plaintiff's forensic expert, served plaintiff well as a consultant during contract performance, and was committed to advancing plaintiff's cause to the extent permitted by the scope of his retention, it was obvious at trial that he faced an uphill battle with numerous constraints hampering the task that plaintiff and the EC/L Project presented him.*

Mr. Bolyard testified as an expert in CPM scheduling for construction projects; delay analysis for construction projects; construction cost estimating; construction cost analysis; and construction means and methods. See supra note 14. Mr. Bolyard has broad experience in the construction industry. He has served as a field engineer, design estimating engineer, and projects engineer for different employers on major construction projects. He has extensive experience in providing cost estimating services; CPM scheduling services; constructability reviews; construction cost analyses, including costs of changes and loss of labor efficiency; and analyses and cost control work. He is fully qualified to monitor and calculate a contractor's performance against a CPM schedule. Significantly, Mr. Bolyard was certified in April 2004 by the AACEI (formerly known by its full name, the Association for the Advancement of Cost Engineering International) as a Planning and Scheduling Professional, which is the only recognized professional certification for practitioners of CPM methodology, and is a member of the American Society of Civil Engineers, the Heavy Construction Contractors' Association, the Construction Management Association of America, and AACEI.

Mr. Midgette was accepted as an expert in the field of scheduling and claims analysis and equitable adjustment claims. Mr. Midgette assists contractors with the resolution of construction contract disputes. He testified that he has experience as a general contractor in residential and light commercial work and that he has served as an office manager, where he maintained and copied drawings, sent them to the field, and maintained necessary data. He also has experience with CPM scheduling, loss of productivity analysis, and recovery schedules. Mr. Midgette is certified by AACEI as a Planning and Scheduling Professional, which notably, however, he received after he submitted his revised expert report of August 30, 2004, as he sat for the AACEI certification exam on October 6, 2004. Tr. at 827. He has never prepared a bid estimate for a project on behalf of a construction contractor. Mr. Midgette has a promising career as a construction consultant, because he showed himself to be dedicated and enthusiastic, but Mr. Bolyard's background, range of experience, and knowledge of the Project overwhelmed Mr. Midgette's delay analysis and proved to be a more reliable source of expert opinion on the issues surrounding the EC/L Project. (emphasis added)"

Finally, MCAA factors are merely the start to an expert's analysis, not the end of his/her work. Sunshine's expert cited a number of MCAA factors in his determination of Sunshine's claim for labor inefficiencies.

However, the court noted with approval the testimony of ACOE's expert, who warned that such factors are not meant to provide absolute costs nor percentages but rather as a general reference, citing the language within MCAA Bulletin 58. Sunshine's expert never tested the reliability of the percentages used by the MCAA against the facts on this jobsite.

To summarize the federal decisions on this point as succinctly as possible, there is not one decision since the *Daubert* trilogy of cases that has permitted expert testimony based primarily or solely upon the MCAA "industry standards."

Federal Administrative Board Decisions Relating to MCAA

In six different cases, the administrative courts have rejected testimony based on MCAA factors, for a variety of fact-specific reasons.

In Appeal of *States Roofing* Corporation, A.S.B.C.A. No. 54860, 10-1 B. C. A. (CCH) 34356, 2010 WL 292732 (Armed Serv. B.C.A. 2010) the Armed Services Board of Contract Appeal (Board) noted other administrative boards had accepted the MCAA analysis, "despite its subjectivity," but further stated it had rejected those same MCAA standards in Appeal of *AEI Pacific, Inc.* A.S.B.C.A. No. 53806, 68-1 B.C.A. (CCH) 33792, 2008 WL 436928 (Armed Serv. B.C.A. 2008) Inc. and *In re Sauer, Inc.* A.S.B.C.A. No. 39605, 61-2 B.C.A. (CCH) 31525, 2001 WL 865382 (Armed Serv. B.C.A. 2001).

In *States Roofing*, the testimony incorporating the MCAA factors, and the resulting estimated labor inefficiency, came from the contractor's President. The use of a standard lacking in objectivity, offered through the testimony of an executive of one of the parties, was too much subjectivity—and thus unreliability—for the court to embrace. The court's decision is further appreciated by the fact that the President testified to a 340% loss of productivity, a figure that strains credulity to the breaking point.

In *AEI Pacific*, the expert was not sufficiently aware of the facts of the case, in that he neither spoke to anyone from AEI prior to the submission of his report, nor did he visit the job site. Instead, the expert based his testimony on the MCAA factors and his review of the party's internal documents. Due to the fact he did not take many notes during that review, the expert could not explain what specific MCAA factors and numbers he applied to the party's documents, at what intervals, and therefore how he calculated the party's alleged inefficiency costs. As a result, the court concluded the expert's opinion, based on MCAA factors but without specific justification and backup data, was "speculative and lacking credibility."

In *Herman B. Taylor v. General Services Administration,* G.S.B.C.A. No. 15421, 01-2 B.C.A. (CCH) 32320, 2003 WL 21711359 (Gen. Services Admin B.C.A. 2003) the MCAA factors were determined to only be appropriately used in calculating the labor inefficiency for mechanical workers, but not for other trades.

In those three cases where the MCAA factors were admitted in part and rejected in part, the facts of each case are critical to the ultimate decision—and thus the decision to permit such expert testimony must be taken with a grain of salt.

For example, in *Turner Construction Company v. Smithsonian Institution,* CBCA No. 2862, 17-1 B.C.A. (CCH) 36739, 2017 WL 1968758 (Civilian B.C.A. 2017) a 2017 decision of the U.S. Civilian Contract Board of Appeals (Board), two of the eight subcontractors, using MCAA factors, sought inefficiency payments from the Smithsonian Institution (Smithsonian) for their work on the renovations to the National Museum of American History. One subcontractor succeeded, while another failed.

At the outset of its lengthy opinion, the Board notes that the *Smithsonian* never agreed to a firm fixed price with the general contractor before construction commenced, and it now had a renovated building. The court's sense of equity was evident on the first page of its opinion, when it states "[h]aving failed to execute the bargain prior to the provision of services, Smithsonian cannot reap the benefit of the bargain it wishes it had struck."

As to the subcontractor that succeeded through its reliance on MCAA factors, the court noted that the use of the measured mile method was not available due to the nature of the work and the nature of the disruptions caused by the Smithsonian's changes. The subcontractor therefore applied four of the MCAA factors to calculate a 42% inefficiency rate. Perhaps even more important, and certainly more persuasively, the subcontractor then reduced its claim of additional or adversely affected hours by 25% to address defects in its own performance before multiplying the inefficiency rate by the number of hours affected. A reader of this opinion could conclude that the subcontractor's method was viewed with favor not because of the accuracy or applicability of the MCAA factors, but because of the equity the subcontractor employed in its calculations—in stark contrast to the positions taken by the Smithsonian.

The court denied the other subcontractor that also utilized the MCAA factors simply because there was no evidentiary support for recovery. Although one of their witnesses testified that the subcontractor had consulted with tables in industry standards, they did not provide that table to the court. This subcontractor did not offer to the court any expert testimony to support its application of the MCAA factors. Without that kind of evidentiary support, any labor inefficiency claim would fail, as this one did.

The case of *In re Fire Systems, Inc.*, V.A.B.C.A. No. 5559-63, V.A.B.C.A. No. 5566-70, V.A.B.C.A. No. 5579, V.A.B.C.A. 5583, V.A.B.C.A. No. 55581, 02-2 B.C.A. (CCH) 31977, 2002 WL 1979118 (Veterans Admin. B.C.A. 2002), was related to the pipe and sprinkler installation of a new fire suppression system at the existing Veterans Hospital in Columbia, Missouri in 1994-5. The VA Board noted that the government had conducted a study and therefore knew prior to the time it requested bids of the presence of asbestos in the building and made those that attended the pre-bid walk-through aware of that fact. But the government did not mention that report in the solicitation for bids, nor was there any amendment to that solicitation. The contractor did not become aware of the potential problem until it encountered possible asbestos in the middle of the job.

The contractor subsequently made a claim for labor inefficiencies, based primarily upon its labor logs and three different MCAA productivity factors. The Board accepted the argument that the discovery of possible asbestos may have had an effect on one factor, morale and attitude, it substantially reduced the risk from "severe" to "minor" because of other facts, including the prompt response by the VA to remediate the situation. The other MCAA factors raised by the contractor were flatly rejected, as there was no support for such claims in the daily logs.

The third case in the last two decades where the MCAA standards were accepted in part and rejected in part was the case of *In re Clark Construction Group, Inc.*, V.A.B.C.A. No. 5674, 00-1 B.C.A. (CCH) 30870, 2000 WL 375542 Veterans Admin. B.C.A. 2000). The subcontractor's inefficiency claim related to the construction of a VA hospital in West Palm Beach, Florida, in the early 1990s. The subcontractor attempted, without success, to use the "measured mile" approach to establish damages, and also argued the MCAA factors in the alternative. Due to the board's finding that the change of construction sequence "pervaded the entire project," and the subcontractor's bid was "reasonable," the board was equitably inclined to render some award for the claimant. Since the "measured mile" was not warranted, the board used two of the MCAA factors—dilution of supervision and site access—to calculate an award, but at a reduced rate than that argued by the subcontractor. The Board chose not to use other MCAA factors because the subcontractor failed to provide relevant evidence supporting the impact on productivity, despite the presence of the subcontractor's voluminous contemporary records. The board specifically made the inference that those records did not justify any other claims, or the claimant would have brought those records to the board's attention.

Industry Standards: National Electrical Contractors of America (NECA)

The admissibility of an expert's testimony based upon NECA industry standards has been the subject of only one published federal court opinion since the *Daubert* standards were adopted, and that was in the *Trane* decision previously mentioned. In that case, Judge Treadwell, in a footnote, dismissed the NECA standards as not meeting the *Daubert* standard in the following way:

> *"The problem with the NECA study is not that its data are inaccurate, but that it considered electrical work, not Mechanical work. So, the MCAA's observation that no one has proved the underlying data inaccurate is not only logically flawed, but also unresponsive to the studies critics."*

There has also been only one state court decision, *Norment Security Group Inc. v. Ohio Department of Rehabilitation and Correction*, 2003-Ohio-6572 2003 WL 22890088 (Ct Cl. Ohio 2003) which also declined to accept the expert's testimony based on MCAA and NECA factors. *Norment* involved the construction of a prison near Toledo, Ohio. The contractor's expert testified that his calculations were based on MCAA factors, and then estimated that inadequate scheduling and coordination caused a productivity loss for the electricians, carpenters, and iron workers on the job. Despite his testimony that his assessment represented an average or medium impact on productivity, the Ohio Court of Claims found the expert's findings based on data from the MCAA manual to be "arbitrary and speculative and do not represent a reliable measure of damages." The Court of Claims also refused to accept NECA factors due to insufficient evidence.

Industry Standards: Business Roundtable (BRT)

In *Trane*, the federal court also rejected the BRT industry standard, writing in a footnote that "defending the accuracy of the data misses the issue: the problem with the Business Roundtable study, for instance, is not that the underlying data are inaccurate, but that the underlying data are limited to one project over twelve weeks."

Other than the *Trane* decision, there has been one federal court decision, the 2006 case of *Ace Construction Inc* v. U.S.,70 Fed. Cl 253 (2006), aff'd 499 F 3d 1357 (Fed Cir 2007) that analyzed this industry standard. In that case, the contractor brought a "differing site condition" claim against the government arising out of inadequate or erroneous representations made during the bid process. The court found that the government's initial drawings were inaccurate, leading to additional labor and costs for the contractor. As a result of those differing site conditions, the contractor argued, and the court agreed, that it was entitled to $462,745 in direct costs (including additional labor costs, additional equipment rental costs, and additional overhead costs of field office expenses, lost profits, and bond expenses). It was also awarded $17,283 for its constructive acceleration claim, due to the fact that the contractor had to incur overtime costs for a considerable period of time to avoid paying liquidated damages as set forth in the contract. Remarkably, the government would not extend the end date of the contract after the government had been presented with evidence of the problems—and even adopted part of the general contractor's proposed solution.

Finally, the court awarded the general contractor its loss of productivity claim for $12,651 in labor costs, plus the same kind of overhead costs mentioned previously. The court had already ruled in favor of the general contractor on all its other claims in the prior sections of its opinion and did so here—except it refused to grant recovery for the equipment inefficiency costs. The expert had taken the Business Roundtable efficiency charts for 50- and 60-hour workweeks, and made adjustments to fit the facts of the case. The court found the BRT to be "relevant" and further noted that the general contractor "*presented uncontroverted evidence* that such overtime resulted in reduced productivity of labor."

This is the only published opinion since *Daubert* where an expert's reliance on the BRT has been permitted. In light of the compelling facts of the case in favor of the contractor, the court's acceptance of all the other contractor's damage claims presented, the fact there was no defense to the use of the BRT tables, and the fact that this inefficiency claim represented only roughly 5% of the direct costs claimed by the general contractor, a party should be forewarned to present expert testimony relying solely on BRT tables.

Industry Standards: United States Army Corps of Engineers (ACOE)

In addition to the aforementioned *Trane* decision, there is only one other reported decision since 2000 assessing the reliability of the ACOE standards, that being the administrative board decision in the *Hensel Phelps* case, which rejected such testimony. One of the reasons there is such a paucity of case law on this particular standard is likely due to the fact that several years ago the Corps of Engineers withdrew its 1979 publication, Modification Impacted Analysis, as the court noted in *Trane*.

Industry Standards: Bureau of Labor Statistics

As set forth in Chapter 6, in 1947 the U.S. Department of Labor published Bulletin 917 regarding inefficiency costs related to the overtime hours for a sole munitions manufacturer during World War II. The application of this industry standard is dubious. In the 2000 ruling of the Corps of Engineers Board, Appeals of J. A. Jones Construction Company, E.N.G.B.C.A. No. 6348 00-2 B.C.A. (CCH) 31000 WL 1014011 (Corps Eng'rs B.C.A. 2000) the Board specifically rejected the study because it only applied to manufacturing plants and did not apply to construction projects. This standard has not since been assessed in any published case.

The Application of the *Daubert* Standard in Construction Cases: The Likely Failure of Future "Industry Standards" Methods

This review of the published court and agency board decisions demonstrates that the purported "industry standards" have fared very poorly since the courts have instituted the higher and more rigorous standards of *Daubert*, *Joiner* and *Kumho Tire*. Some future "experts," in support of their future theories, may try to explain this consistent rejection of current industry standards as simply due to the fact that these studies are very old and/or the underlying data is no longer subject to peer review. Such claims would simply not be true.

As reflected in the cases reviewed in this chapter, the age of these studies is not the only—or even main—reason why the industry standards have fared so poorly. For all the reasons set forth in Chapter 8, these so-called "industry standards" are not directly applicable to the facts of any case precisely because each construction project is different from any other—often in multiple ways. Any future proposed "industry standards" will likely fail, and should be disregarded, for the very same reason: there are too many variables in any and every construction project to believe in a method that is not based on the facts of each construction project.

In other words, the most reliable method of the future is the most reliable method of the past and present: the measured mile. The only way to be able to employ the measured mile is to have assembled facts that reasonably compare the productivity of the work performed before the project was impacted with the productivity of the work that was performed after the project was impacted. And the only way to have meaningful comparative data is to be vigilant and dedicated to generating that kind of data as the job is progressing—not attempting to recreate it after the completion of the project. This approach not only pays dividends in increased profits during the construction process, but it also benefits the parties should a dispute arise that requires such documentation as factual evidence. In addition, this approach provides the data that will serve as the basis for the expert witness' opinion.

12

Achieving the Desired Results

The authors recognize that events frequently arise in construction projects that could adversely impact the productivity of the workforce. To best respond to that event, the parties must first identify there is a problem, and agree it can have a potentially negative effect on the project's scope, schedule and/or budget.

Once the problem has been identified, we have recommended that the parties work together to solve "their" collective problem. We urge the parties to reject the finger-pointing approach of the past, an approach that has led to so much wasted time and money. We have instead promoted the concept of problem-solving collaboration, a concept we have seen work well on certain projects.

To be able to identify such problematic events in real time, the parties should collect meaningful and accurate labor-and work-related data (preferably on a daily basis), evaluate that data on a frequent basis, and then hold people accountable for such accurate and timely data generation and the performance of the work.

We have also recommended that a party retain experienced construction counsel (and possibly an experienced construction expert) early in the problem-solving process, with the initial goal of monitoring and addressing the disruptive event as quickly and as inexpensively as possible. Should it be difficult for the parties to solve the problem while construction is underway, or agree on the responsibility or cost of such problem's solution, at least the parties will have narrowed the issues of disputed fact and placed themselves in a better position to more quickly settle their dispute at a later date.

It is well recognized that most disputes are settled prior to the parties submitting their case to the trier of fact, be it a judge, jury, administrative judge, or arbitration panel. More and more of those settlements now occur as part of a mediation process, usually after expert witness testimony has been obtained by the opposing party. With these unmistakable trends in mind, what is the best way to obtain the most favorable settlement terms?

Ted Trauner, Chris Kay and Brian Furniss. Productivity in Construction Projects, (191–196) © 2022 Scrivener Publishing LLC

We offer the following suggestions:

First and foremost, make it a priority to retain qualified personnel to collect relevant data and review it frequently - from the commencement of the initial budget and schedule to the completion of the project. These people can not only keep the job on schedule and on budget, but they will likely become important fact witnesses should litigation or arbitration be necessary. With the sophistication of modern project control software programs that monitor various aspects of the project, the retention of knowledgeable project staff can also offer crucial support to a party's expert, by providing them with accurate and relevant information immediately after the expert has been retained. We stress that the retention of such support personnel is not an extra expense to the party. To the contrary, they can save the party money while the job is ongoing by working to decrease inefficiencies, and they can help the party recover money (or avoid the payment of money) if litigation or arbitration is necessary.

Second, retain an experienced attorney—someone who has proven that he/she knows there are ways to avoid unnecessary time and expense in the resolution of construction disputes. Select someone who is not trying to make a fortune on the legal fees that will be generated in this particular case. A party should insist that their attorney actually performs a substantial amount of the critical work and/or be significantly engaged in the assessment of liability and damages. Don't let the experienced attorney, once retained, subsequently delegate the vast majority of the work to a younger associate - all in the purported context of "saving the client money." To the extent that courts are not permitting the testimony of experts who are not intimately knowledgeable about the facts of the case, a party should not hire an attorney to argue its case unless that attorney can meet that same standard.

Third, have a party's attorney retain an expert (or experts) that know the construction business "from the inside"—people who are contractors, schedulers, and/or engineers. That expert also must be committed to learning all the relevant facts of the construction project. As we have seen from several of the published court decisions, a witness who simply teaches at a university, relies heavily upon "industry standards," and/or computes damages based solely upon overtime and wage data is not going to meet the *Daubert* standard for expert testimony.

Fourth, a party should insist that its attorney and expert work together to assess the critical factual issues, the possible solutions to the current construction problems, and the evidence proving (or disproving) labor inefficiencies. This approach creates several benefits to the party. First, such efforts may help solve the problem while the project is ongoing, thus minimizing the disruption. Second, to the extent there is subsequent litigation or arbitration, their initial work should result in discovery conducted with a laser-like focus on the most relevant facts. That kind of discovery should save time and money. Finally, when it comes time for the mediation, the attorney and the expert should be able to collaborate to make the most effective presentation possible.

Fifth, a party should push hard for an early mediation. The purpose of hiring experienced and engaged attorneys and experts, and then insisting they conduct discovery quickly, is to push for a mediation as soon as possible. The opposing party may object to this approach, but the party moving for mediation can make the argument that (a) the case has been pending for a considerable period of time; (b) that same party has followed the court's orders (or those of the arbitration panel) and diligently conducted its discovery; and (c) that party is now ready to have the matter resolved, either in mediation or at trial. The judge or panel may give the other side some additional time for discovery, but generally the additional time is likely not enough to overcome the advantage an aggressive party's attorney and expert will have gained by working diligently to that point. This approach will generally result in an earlier mediation date, and less legal expenses associated with discovery, than had the party let the typical legal process dictate the date of mediation. Preparation is critical to success, and a party's attorney and expert should work hard to be better prepared than their adversaries, and to do so as quickly as possible.

Sixth, the senior-most decision makers for the parties need to be present when the time comes for the mediation. A party's owner (assuming he/she is the CEO or President of the party) should make it clear that he/she will attend and have its attorney call for the opposing party's CEO or President to also attend. Parties are generally required by the court to bring a "decision maker" to the mediation. In many cases, that person is the company's General Counsel. The reason to have an opposing party's owner or CEO attend, particularly if that owner or CEO is experienced in construction work, is that the executive will see the evidence as it would be presented to the trier of fact—not filtered through the subsequent report from that party's General Counsel. A senior construction executive also has likely seen a lot of "real life" situations in the field similar to those described in the mediation and may have a greater appreciation for the fact that no party is ever completely blameless or completely to blame.

Seventh, if a party is able to persuade the opposing party's CEO or President to attend, careful consideration should be given to also suggesting the experts participate in the mediation. Seasoned construction executives may tend to discount some of the arguments made by opposing counsel, on the grounds that lawyers get paid to advocate for their clients. But an objective expert, with considerable experience working for or in companies like that of the opposing party, may have an opportunity to make a more persuasive factual argument than an attorney—and thus be taken more seriously by the opposing party's CEO.

Case in point: The authors engaged in such a settlement session, after the owner had terminated the general contractor and sued it for $10 million in federal court. The general contractor filed a $30 million counter-claim. The parties agreed to bring their most senior executives, attorneys, and experts together. The general contractor was a large, national company—and its CEO attended the session. Over the course of an entire day, the owner's expert explained his opinions, supported by sound construction and scheduling principles, visual proof, and construction documents. The dispute was settled the following day, on terms favorable to the owner. The authors believe that the general contractor's CEO could not have fully appreciated the evidence against his company had he not been in attendance at that session.

What is the bottom line of a construction dispute? It is the amount that one party pays, and another party receives—it is the parties' financial bottom line. The authors appreciate the fact that events occur during the construction process that can result in a loss of productivity and an increase in labor costs. This is a business problem not uncommon in the industry. Those that have successfully solved that business problem during or immediately after the completion of the project tend to spend less money on attorneys and experts and save more of their valuable time. To achieve that result, seriously consider retaining experienced personnel early in the process to find greater efficiencies and/or document and address inefficiencies. Those professionals can certainly help save money during the construction and/or thereafter in the litigation or arbitration.

Should a party become involved in litigation or arbitration, retain attorneys that recognize they are engaged in but one way to solve a business problem—usually by settlement. Hire professionals who fully appreciate that the bottom line for their client's dispute really is the client's bottom-line cost for all the expenses associated with the litigation or arbitration. An attorney should not waste the client's money retaining an expert witness that will not be certified as an expert under the *Daubert* standard. Neither the party, the attorney nor the expert should waste time and money by letting the dispute get dragged out for years of discovery, the all-too-common approach for many construction cases today. Instead, everyone's focus should be on the best way to bring the parties to a mediation or settlement conference as soon as possible.

It is the authors' experience that the amount paid in settlement for alleged inefficiencies is likely going to be the same if a case is settled right before trial or a year before trial—since the dispute is based on the facts of the project, and the expert's analysis of those facts. The only difference between an earlier settlement and a later settlement is the additional legal expenses incurred over that additional year. We have been involved in disputes where a party grits its teeth and pays a higher sum than it initially thought was warranted, but did so because that party's legal fees were exacting a substantial toll. Consequently, we recommend swiftly conducting focused discovery and moving for a quicker mediation, precisely so that a party can—if necessary—make a larger payment than originally intended but achieve a resolution to the dispute. That type of settlement avoids another year of expensive, and likely not beneficial, discovery. That additional year of discovery may also distract the parties from spending more of their time in pursuit of their business endeavors.

13
The Way Forward

Collectively, the authors have over 120 years of experience in the construction industry. However, at no point have we waxed poetic about the "good ol' days." We did not do so precisely because the "good ol' days" were not "good" for many in the construction business.

The "good ol' days" were not good for those organizations that did not know the most basic and integral elements of labor productivity, or lack thereof, and thus bid their jobs without the most important information. Those companies got hammered as the job proceeded. In addition, the "good ol' days" were not good for organizations that did not intimately know how to monitor the progress on the job using those elements. As a result, they were not able to make timely adjustments in response to problems that arose on the job. Accordingly, many such parties ended up in litigation—an expensive process that takes a lot of time and effort from more profitable ventures, and seldom makes a party "whole" upon its completion. Finally, the "good ol' days" were not good for organizations that turned to attorneys and expert witnesses that failed to understand the crucial elements of labor productivity on that particular project, instead taking the easier path of arguing "industry standards".

As a result of those mistakes, too many companies were ultimately forced to close their doors, including reputable organizations that had been in business for many decades. We know this to be true, because we ran businesses, or were retained by others in the construction business, that competed against such organizations—and succeeded at their expense.

To succeed then—and to succeed now—people in the construction industry must do three things, whether they are an owner, a general contractor, a subcontractor, an engineer, a construction manager, or even an attorney or expert witness.

First, they must take the time, and make the effort, to understand the basic elements of labor utilization and incorporate that data into the original bid and schedule. You must think in terms of the labor it takes to lay X number of feet of pipe per man-hour, and then take it to the next level by

Ted Trauner, Chris Kay and Brian Furniss. Productivity in Construction Projects , (197–204) © 2022 Scrivener Publishing LLC projecting the productivity if employee X or employee Y is serving as the superintendent or foreman on that part of the job. In these post-COVID days, there is also the question the talent pool. Not only do we need to project the employees X and Y serving as supervisors, but all the employees working for those supervisors. You also must think in terms of the labor it takes to run Z feet of wire at a height of 10 feet per man-hour on the first floor, and then calculate (and adjust) the same costs per man-hour for the same wiring on the next 12 floors, including the time it takes to get workers and equipment to and from those elevated floors.

Second, once the job has commenced, we stress that the parties should periodically review the daily records together. These data review sessions create an opportunity for all the parties to measure labor utilization in the same terms set forth in the original schedule and determine if those projections are still being met. From our decades of experience, we know that major construction jobs seldom always run on time and on budget as initially foreseen, even if there are no changes to the original plans and scope of work. Instead, the frequent monitoring of the data will provide companies with the ability to respond more quickly—and more accurately—to the unexpected issues that arise on that particular job. Such efforts also have a long-term benefit, in that the data will add to the companies' knowledge base for a better educated approach to the next project.

Third, such detailed data review sessions result in greater collaboration, transparency, and accountability throughout the construction process. That is the kind of cultural change of which we write in this book. When all the parties are together reviewing the same detailed labor utilization data in real time, the insights and experience of the different professionals involved can create better, more comprehensive, and immediate solutions. When the parties see all the relevant data, agree upon a modification of the labor utilization, and then check back a few days or weeks later to see if the modification has achieved its desired result, the parties are working together for their collective good.

While the project is under construction, it is much easier to assume the best intentions of everyone involved, precisely because all the parties have the best of intentions: to get the job done right, on time and on budget. But it is our experience that once the project has been completed, often after a series of contentious and accusatory conversations and emails, the parties retreat to their respective corners and prepare for battle in court. That is the wrong culture at any time, and particularly today when there are so many more demands on every player in the construction industry.

We recommend everyone involved in the construction industry spend more time and effort in the planning stages, and more time and effort monitoring the labor utilization of all the parties throughout the subsequent construction phases. We understand that this means more work in the planning stages, and different work than may have been performed in the past. Some executives or employees may not want to learn or utilize this approach, or that they may not want to conduct this kind of work on a daily basis.

We also understand that change, of any kind, can be a pain in the neck. But we have witnessed over our combined careers in various roles in the construction process, time and time again, that too many businesses that didn't embrace change subsequently suffered mightily—and were ultimately forced to close their doors. Which raises the obvious question: Would you rather experience a pain in the neck while your organization learns a different way to measure and monitor labor utilization, or would you rather have your business decapitated?

With respect to our first point, to fully understand and break down the minute aspects of the work and calculate them in terms of function accomplished per man-hour, we are recommending those in the construction business follow the lead of their peers in other industries. In the "good ol' days," we might travel to the airport on roads where we were forced to stop and interact with an operator in a tollbooth. Today, our license plates are captured by cameras that automatically debit our accounts, as we speed by without interruption to our destination. When we arrived at the airport, we used to be greeted by a large group of people at the check-in counter. Today, we obtain our ticket via a kiosk. Technology is at work to enhance productivity, and reduce labor costs, all around us.

Let's look at another industry that captures the passion and imagination of millions of people around the world: sports. The sports world has seen an explosion in the cost of its labor, as salaries of sports stars has skyrocketed to record heights. Some players make so much money during their careers that they become owners after they have retired from their playing days. With that kind of cost for a sports team's labor, those teams have spent incredible time, talent, and money on.... labor analytics. Every sport has broken down the elements of the work of its players into many distinct units that affect the performance of that player, and the success of that team.

Today, fans have twice as many ways to measure the performance of their favorite players. Moreover, that data is made available to them on the internet and through the media with instant updates. If you are one of those sports fans, you are now using constantly updated data analytics—labor productivity metrics—to know and assess the performance of your teams and their players more thoroughly. This begs the next question: If you are using data analytics to measure the performance of the players on your favorite team, why aren't you employing the same data analytics for your construction projects?

Once you have embraced the concept of measuring the productivity of your personnel through labor analytics, the next crucial step is to ensure that you are using the right metrics. That requires getting involved in the "nitty gritty" of each aspect of the work, and then analyzing it a couple of different ways, before determining the working units that will be used to build the bid and schedule, and then measure the work thereafter. We understand that contractors and subcontractors are often bidding competitively on construction jobs. To win the job they must sharpen their pencils and have the best price. What could be a better estimate/bid than one that is based on hard records of demonstrated performance? Obviously, none. But the accuracy of our bidding can be moderated, depending on the scale we can comfortably live with. Allow us to clarify that a bit.

A bidder can use a "market" bidding manual that will provide numbers of specific items for respective areas. Most likely this is not the common route taken. More commonly, the estimate will be based on historical performance data that substantiates actual performance in prior, similar projects. But this too has a variable scale. If the previous jobs demonstrated that a vertical project was constructed for a certain cost per square foot, is that enough to use it as an estimate basis on the current project? Maybe so. But besides knowing the cost per square foot from preceding jobs, would the estimate accuracy increase if the historical cost for each unit of work was used instead of a square foot cost? Probably so. Therefore, a fine tuning of the estimate can occur when established historical records exist that support the costs for each individual unit of work.

If those records are available, is it enough? Maybe not. Hopefully, the records are detailed enough that the cost per unit can be separated for superintendent or foreman because the same people cannot lead and manage every job. Those performance numbers often vary for different participants.

The same reasoning applies for subcontractors. A contractor may have historical data on what a subcontractor costs for certain units of work, but what staff from the subcontractor was supplied on those projects? A subcontractor may have data on what its costs were for units of work—but what superintendent or foremen were on those jobs?

By now, you should see where this discussion is headed. The obvious conclusion is that the better our historical records and data are, the more accurately we can prepare a bid.

Our second overarching recommendation is that once the job has commenced, we stress that the parties should periodically review the daily records together. Once the project starts, the information must be tracked for what is happening on the project, to determine if the actual performance is meeting the estimated performance. If performance is better or worse, the contractor needs to answer the question of "why" and then act.

We understand that we are saying that you need to expend a lot more effort on recordkeeping. And not only must you keep more detailed records, but you must also review and analyze that data to be able to make adjustments along the way and for future work.

If you reflect on this for a bit, in today's tech-based world, we have devices that will assist us in more thorough and more detailed recordkeeping, and then those devices will provide almost instantaneous analysis and performance data. Whether we consider laptops, tablets, or phones—we have the technology and the means to do this work much faster than in the past. Granted, you are going to have to perform some "up front" work, but it is an investment in your company's success. Let's consider that for a moment.

What information do you want to know? What form or format should be utilized to record that information? What level of detail do you need the information to be recorded in? What happens to the information once we collect it? Who is the record keeper or keepers? How often is the daily information collected? How often is that information analyzed, reviewed, assessed? Who does the review or assessment? Bear in mind, that you may record detailed information and may recognize from your periodic review that you are not achieving the level of productivity planned. But without that investment of up-front work, you are unable to ask the important question that stems from that assessment—"WHY"? Were the working conditions different than what we anticipated? Do different foremen achieve markedly different results? Is productivity better or worse than planned? If the level of productivity continues, will there be any effect on the project schedule? Obviously, depending on the results of our demonstrated performance, tasks may finish in less or more time than planned or allowed by the schedule.

We recognize that this increased knowledge may be a double-edged sword. As contractors, if we have more precise historical information, we may not win the bid. But is this really bad? Do we want to win the project bid if we are going to lose money? Over the years, we have heard most of the counters to this position. Some of them include:

1. We can make up any difference on the change orders.
2. We will assign our best people to the job, and they have always succeeded – even with the problem jobs.
3. We will find "opportunities" and costs so we end up profitable.

4. We must take these risks to have a chance at growing the company.

The list goes on.

How about the construction owner? Why should that entity keep detailed records? What possible benefit could an owner receive from having their inspection personnel keep such information? Hopefully, from the examples that have been given throughout this book, an owner should be able to see the benefit he/she can receive. Information on the productivity that is being achieved on a project is directly related to the schedule on the project. Does any owner want to learn that the only way the completion date can be met is if overtime is undertaken? Does an owner want to learn late in the game that the schedule will not be met? Does that owner want to suffer the reputational damage, in the eyes of their customers, if they have spent tons of money and time marketing the grand opening of their new facility, only to learn later that the grand opening will not take place at the promised time? Does an owner want to get embroiled in a costly dispute when the project finishes late or when a claim is submitted asserting that changes by the owner caused significant extra costs in terms of delays or losses in productivity? Of course not!

The owner should be assessing the records and comparing the demonstrated productivity with the durations shown in the project schedule. Candidly, it is very easy to have a project schedule reflect that a project will finish on time even if it can't. It is equally important for the owner to know what actually is being achieved on the project in the same detail. It is equally important for the owner to record, maintain, and assess detailed information on the productivity of various trades on the project so he/she understands the probabilities for a successful project schedule.

Which leads us to our third recommendation, that such detailed data review sessions among the owner, contractors and subcontractors result in greater collaboration, transparency, and accountability throughout the construction process. That is the kind of cultural change which is needed to succeed today. It is also the kind of cultural change that is going on elsewhere in our country's businesses. The whole idea of blockchain technology is to create a permanent, transparent ledger system for the participating parties in that system to use today's technology to compile data on the most relevant aspects of the parties' business. This innovative approach has been successfully used by so many different players in so many diverse industries. Accordingly, it is essential that all parties to a construction contract not only agree to share such data, but they commit to regularly evaluate the data and, when necessary, make changes to increase productivity.

It is important to have collaboration and transparency. But it is equally important to have accountability. Once the parties have agreed to make a change, whether it be to the architectural design, to the work schedule (as in the authorization of overtime), or to such tangential aspects such as the workers' transportation to the job site (as in adjacent parking or addition of elevators), the newly agreed upon modification must be periodically reviewed to determine if it is achieving the desired results. A change in plans, without accountability, serves no beneficial purpose to anyone.

Finally, to ensure accountability, it is imperative that the information shared and reviewed by all the parties is accurate. Accuracy is a function of several important players in the process. It starts with those responsible for the elements of the bid, which become the units by which the parties measure the productivity of the labor force throughout the project. As the project commences, different people, such as foremen or construction managers, are responsible for generating progress reports. It is essential that these reports not only be accurate, but measure productivity with the same units of work performed that were set forth in the schedule in place at the commencement of the project. Otherwise, how can the parties truly know the state of the project's progress, or lack thereof, and make the necessary changes?

But it is not enough to rely solely upon the honesty and accuracy of the foremen and construction managers because those individuals may be knowingly or unknowingly influenced by their bosses, or their bosses' bosses. We have seen it happen too many times, where a work progress assessment is "modified" or there is a schedule "update" that makes it appear that the project is on schedule—when in fact it is not. In such instances, the owner (and perhaps others) is temporarily lulled into a false sense that everything is proceeding according to plan. But the day of reckoning eventually occurs, and those that thought they could surreptitiously recover lost time and productivity are unable to do so. The project's completion is delayed, to the deep disappointment and financial loss to many involved.

In conclusion, all the parties to a construction project should take the time, and make the effort, to understand the basic elements of labor utilization and incorporate that data into the initial bid and schedule. Once the job has commenced, we stress that the parties should periodically review the daily records together. Finally, to ensure accountability, it is imperative that the information reviewed by all the parties is accurate. Based on our experience, this is the best way to enhance labor productivity, avoid labor inefficiencies, and manage the project through all its issues to a successful completion.

Whatever additional work that may be required for your organization to change in this fashion, we believe it will be well worth your time and effort—and consistent with the same kind of changes that have already taken place in other industries. Greater labor productivity results in greater profitability and a greater likelihood of greater business opportunities. It has become the way forward for many other industries in this country. We have seen this approach work successfully on some of our projects. We hope you will make this approach your way forward as well.

We understand that we are strongly suggesting that all parties in a construction project spend more time and money tracking, documenting, and analyzing the detailed work on the project. Yes, we are! And we have witnessed that effort pay for itself and generate success and reduced costs later.

Bibliography

"Absenteeism and Turnover," Report No. C-6, The Business Roundtable, New York, NY 1982.

"Administration and Enforcement of Building Codes and Regulations," Report No. E-1, The Business Roundtable, New York, NY, October 1982.

"Ahuja, II.N. and V. Nadakumar, "Simulation Model to Forecast Project Completion Time," Journal of Construction Engineering and Management, American Society of Civil Engineers, Volume 111, No. 4, December 1985.

Alfred, L.E., Construction Productivity, McGraw-Hill Book Company, New York, NY, 1988.

Bianco, J.T., "Estimating Productivity," Paper presented to the Utility Cost Management Committee Workshop of the American Association of Cost Engineers in Newport Beach, CA, February 1983.

Borcherding, J.D. and D.F. Garner, "Workforce Motivation and Productivity on Large Jobs," Journal of the Construction Division, American Society of Civil Engineers, Volume 107, No. C03, September 1981.

Brauer, R.L., G.J. Brown, E. Koehn, S.T. Brooks, and T. Mahon, "AFCS Climactic Zone Labor Adjustment Factors," Technical Report P-165, U.S. Army Corps of Engineers, August 1984.

Budwani, R.N. "The Data Base for U.S. Power Plants," Power Engineering, January 1985.

"Change Orders," Bulletin CO 1, Mechanical Contractors Association of America (MCA), Rockville, MD, 1994.

"Contractual Arrangements," Report No. A-7, The Business Roundtable, New York, NY, October 1982.

"Construction Labor Motivation," Report No. A-2, The Business Roundtable, New York, NY, August 1982.

"Construction Technology Needs and Priorities," Report No. B-3, The Business Roundtable, New York, NY, August 1982.

"Cost Estimating, Budgeting, and Control Accounting," Appendix A-6.2 to Report A-6, The Business Roundtable, New York, NY, March 1983.

"Cremeans, J.E., "Productivity in the Construction Industry," Construction Review, May/June 1981.

Edmondson, C.H., "You Can Predict Construction-Labor Productivity," Hydrocarbon Processing, July 1974.

"The Effect of Multi-Story Buildings on Productivity," National Electrical Contractors Association, Inc. (NECA), Bethesda, MD, 1975.

"Effect of Scheduled Overtime on Construction Projects," American Association of Civil Engineers (AACE) Bulletin, Volume 15, No. 5, October 1973.

"The Effects of Scheduled Overtime and Shift Schedule on Construction Craft Productivity," Construction Industry Institute, Source Document 43, The University of Texas at Austin, Austin, TX, December 1988.

"The Effect of Temperature on Productivity," National Electrical Contractors Association (NECA), Washington, D.C., 1974.

Emmons, M.W., "Project Management Starts Before Contract Award," Proceedings, Project Management Institute, 1978.

"Flexible Hours Little-Used," Engineering New Record, August 1981.

Gates, M. and A. Scarpa, "Optimum Number of Crews," Journal of the Construction Division, American Society of Civil Engineers, Volume 104, No. CO2, June 1978.

Grimm, C.T. and N.K. Wagner, "Weather Effects on Mason Productivity," Journal of the Construction Division, American Society of Civil Engineers, Volume 100, No. CO3, September 1974.

"Guide to Electrical Claims Management," National Electrical Contractors Association, Inc. (NECA), Volume 1, Second Edition, Bethesda, MD, 1985.

Handa, V.K. and D. Rivers, "Downgrading Construction Incidents," Journal of Construction Engineering and Management, American Society of Civil Engineers, Volume 109, No. 2, June 1983.

Hoffman, P.S., Editor, Law Office Economics & Management, Volume XXV, No. 1, Callaghan & Company, Wilmette, IL, Spring 1984.

"Hours of Work and Output," U.S. Department of Labor, Bureau of Labor Statistics, Bulletin No. 917, Washington, D.C.

"How Much Does Overtime Really Cost," Bulletin OT 1, Mechanical Contractors Association of America (MCA), Rockville, MD, 1994.

Ibbs, C.W. and Allen, W.E., "Quantitative Impacts of Project Change," Construction Industry Institute, Source Document 108 to Publication 43-2, May 1995.

Ibbs, William, "Impact of Change's Timing on Labor Productivity," American Society of Civil Engineers, Journal of Construction Engineering and Management, November 2005, pages 1219 through 1223.

Ibbs, William and McEniry, Gerald, "Evaluating the Cumulative Impact of Changes on Labor Productivity – an Evolving Discussion," AACE's *Cost Engineering*, Volume 50, Number 12, December 12, 2008, pages 23 through 29.

Ibbs, William and Vaughan, Caroline, "Change and the Loss of Productivity in Construction: A Field Guide," Version Date: February 2015.

"Integrating Construction Resources and Technology into Engineering," Report B-1, The Business Roundtable, New York, NY, August 1982.

Jansma, G.L., "The Relationship Between Project Manning Levels and Craft Productivity for Nuclear Power Construction," Project Management Journal, Volume 19, February 1988.

Jansma, G.L., "A Methodology for Making Construction Productivity Comparisons," Doctoral Dissertation presented at the University of Texas at Austin, Austin, TX August 1987.

Jones, L.R., "Overtime-Pay Problems," Chemical Engineering, June 28, 1971.

Koehn, E. and G. Brown, "Climate Effects on Construction," Journal of Construction Engineering and Management, American Society of Civil Engineers, Volume 111, No. 2, June 1985.

Kuipers, E.J., "A Method of Forecasting the Efficiency of Construction Labor in Any Climatological Conditions," Thesis presented at the University of Illinois at Urbana-Champaign, Urbana, IL, 1976.

"Labor Productivity," Mechanical Contractors Association of America (MCA), Rockville, MD, 1994.

"Labor Productivity Adjustment Factors," Prepared for U.S. Nuclear Regulatory Commission, Science and Engineering Associates, Inc., Albuquerque, NM, March 1986.

Leonard, C.A., "The Effect of Change Orders on Productivity," The Revay Report, Revay and Associates Limited, Volume 6, No. 2, August 1987.

Leonard, C.A., "The Effects of Change Orders on Productivity," Thesis Presented at Concordia University, Montreal, Quebec, Canada, August 1988.

Li, S., "Basic Construction Principles in Higher Latitudes," Journal of the Construction Division, American Society of Civil Engineers, Volume 100, No. CO1, March 1974.

Logcher, R.D. and W.W. Collins, "Management Impacts on Labor Productivity," Journal of the Construction Division, American Society of Civil Engineers, Volume 104, No. CO4, December 1978.

MacAuley, P.H., "Economic Trends in the Construction Industry, 1965-80," Construction Review, U.S. Department of Commerce, May-June 1981.

"Materials Management," Appendix A-6.5 to Report A-6, The Business Roundtable, New York, NY, February 1983.

McGlaun, W., "Overtime in Construction," American Association of Civil Engineers (AACE) Bulletin, Volume 15, No. 5, October 1973.

"Measuring Productivity in Construction," A-1 Report of the Construction Industry Cost Effectiveness (CICE) Project, The Business Roundtable, New York, NY, September 1982.

"Measuring Productivity in Construction," A-1 Report of the Construction Industry Cost Effectiveness (CICE) Project, Updated Reprint, The Business Roundtable, New York, NY, 1989.

Mitchell, T.R. and J.R. Larson, Jr., People in Organizations; An Introduction to Organizational Behavior, 3rd Edition, McGraw-Hill, 1987.

"Modification Impact Evaluation Guide," Department of the Army, Office of the Chief of Engineers, Washington, D.C., July 1979.

"Modifications and Claims Guide," Pamphlet No. 415-1-2, Department of the Army, Office of the Chief of Engineers, Washington, D.C., October 1976.

"Modifications and Claims Guide," Report EP 415-1-2, U.S. Army Corps of Engineers, Office of the Chief of Engineers, Washington, D.C., July 1987.
"Modern Management Systems," Report No. A-6, The Business Roundtable, New York, NY, November 1982.
"More Construction For the Money," Summary Report of The Construction Industry Cost Effectiveness Project, The Business Roundtable, New York, NY, January 1983.
"More Construction For the Money," Summary Report of The Construction Industry Cost Effectiveness Project, The Business Roundtable, New York, NY, July 1990.
"Network Analysis Systems Guide," Report EP 415-1-4, U.S. Army Corps of Engineers, Office of the Chief of Engineers, Washington, D.C., 1986.
O'Connor, J.T., "Impacts of Constructability Improvement," Journal of Construction Engineering and Management, American Society of Civil Engineers, Volume 111, No. 4, December 1985.
O'Connor, J.T., M.A. Larimore, and R.L. Tucker, "Collecting Constructability Improvement Ideas," Journal of Construction Engineering, American Society of Civil Engineers, Volume 112, No. 4, December 1986.
O'Connor, J.T. and R.L. Tucker, "Industrial Project Constructability Improvement," Journal of Construction Engineering and Management, American Society of Civil Engineers, Volume 112, No. 1, March 1986.
O'Connor, L.V., "Overcoming the Problems of Scheduling on Large Central Station Boilers," Proceedings of the American Power Conference, Volume 31, 1969.
Oglesby, C., H. Parker, and G. Howell, Productivity Improvement in Construction, McGraw-Hill Book Company, New York, NY, 1989.
"Overtime and Productivity in Electrical Construction," National Electrical Contractors Association (NECA), Washington, D.C., 1969.
"Overtime and Productivity in Electrical Construction," National Electrical Contractors Association (NECA), Washington, D.C., 1989.
"Planning and Scheduling," Appendix A-6.1 to Report A-6, The Business Roundtable, New York, NY, February 1983.
Sanvido, V.E., "Conceptual Construction Process Model," Journal of Construction Engineering and Management, American Society of Civil Engineers, Volume 114, No. 2, June 1988.
"Scheduled Overtime Effect on Construction Projects," Report No. C-2, The Business Roundtable, New York, NY, 1980.
"Scheduled Overtime Effect on Construction Projects," Report No. C-2, The Business Roundtable, New York, NY, 1989.
Scherer, J.V. and R.L. Tucker, "The CII Model Plant," A Report to the Construction Industry Institute, the University of Texas at Austin, Austin, TX, December 1986.
Suhanic, G., "Change Orders Impact on Construction Cost and Schedule," GSCON Ltd., Ontario, Canada.

"Technological Change and Its Labor Impact on Four Industries," Bulletin 2316, U.S. Department of Labor, Bureau of Labor Statistics, December 1988.

Thomas, H.R. and D.A. Anderson, "A Procedure for Evaluating the Efficiency of Power Operated Cutting Tools in Localized Pavement Repair," Transportation Research Record, Transportation Research Board, 1981.

Thomas, H.R. and D.F. Kramer, "The Manual of Construction Productivity Measurement and Performance Evaluation," Report to the Construction Industry Institute (CII), Austin, TX, May 1988.

Thomas, H.R. and C.T. Mathews, "An Analysis of the Methods of Measuring Construction Productivity," Report to the Construction Industry Institute (CII), 1986.

Thomas, H.R., C.T. Mathews, and J.G. Ward, "Learning Curve Models of Construction Productivity," Journal of Construction Engineering and Management, American Society of Civil Engineers, Volume 113, No. CO4, December 1987.

Thomas, H.R. and K.A. Raynar, "Effect of Scheduled Overtime on Labor Productivity: A Quantitative Analysis," Report SD-98, The Construction Industry Institute, The Pennsylvania State University, University Park, PA, August 1994.

Thomas, H.R. and S.R. Sanders, "Procedures Manual for Collecting Productivity Related Data of Labor-Intensive Activities on Commercial Construction Projects," Revision 1, The Pennsylvania State University, University Park, PA, May 1987.

Thomas, H.R., V.E. Sanvido, and S.R. Saunders, "Impact of Material Management on Productivity - A Case Study," Journal of Construction Engineering and Management, American Society of Civil Engineers, Volume 115, No. 3, September 1989.

Thomas, H.R. and I. Yiakoumis, "Factor Model of Construction Productivity," Journal of Construction Engineering and Management, American Society of Civil Engineers, Volume 113, No. CO4, December 1987.

Thomas, Jr., H.R. and G.R. Smith, "Loss of Construction Labor Productivity Due to Inefficiencies and Disruptions: The Weight of Expert Opinion," The Pennsylvania Transportation Institute, Report 9019, The Pennsylvania State University, University Park, PA, December 1990.

Thomas, Jr., H.R., J.R. Jones, Jr., W.T. Hester, and P.A. Logan, "Comparative Analysis of Time and Schedule Performance on Highway Construction Projects Involving Contract Claims," Draft Final Report No. DTFH61-83-C-00031, Prepared for the Federal Highway Administration, July 1985.

"Work Sampling and Foreman Delay Surveys" Appendix B-3, Supplement to Report A-1, The Business Roundtable, New York, NY, 1982.

About the Authors

Ted Trauner has had the opportunity to author and coauthor several other texts on construction related topics. As a recognized expert in the areas of construction management and disputes on construction projects, Ted wanted to chronicle what has been learned, research other available information, and disseminate this knowledge to the industry. The goal is to advance the general body of knowledge so that the construction industry can improve. The areas of Management, Scheduling, and Delays have been addressed in some of the other writings and teachings. Having seen the carnage arising out of the way parties to the construction process dealt with these issues in the past, Ted believes that the area of Productivity in Construction should be the focus of the construction industry's future efforts. This book represents that effort.

Chris Kay was a trial lawyer for over 23 years and handled a wide range of construction disputes. As a construction litigator, he has extensive experience identifying the data and expert testimony needed to succeed in cases of all sizes and complexity. He also appreciates that a "win" needs to take into account the amount of time the client devotes to the litigation rather than to its normal business, and the cost of legal and expert fees. Thereafter, Chris became the first General Counsel in the history of Toys 'R' Us, and later its Chief Operating Officer. He was responsible for construction of hundreds of new stores, and renovation projects for existing stores, for Toys 'R' Us, Babies 'R' Us and Kids 'R' Us across the country. Chris later served as Chief Executive Officer for the New York State Racing Association and spearheaded a number of capital improvement projects at that organization's three racetracks. As the owner of retail stores, malls, and sporting and entertainment venues, Chris knows that the owner seeks to have the project completed on time and on budget, but not at the cost of the contractor taking detrimental "shortcuts" that may adversely affect the short-term operation or long-term viability of the project.

Brian Furniss is a licensed Professional Engineer in Florida, Texas and Colorado, and a Planning and Scheduling Professional (PSP) and Certified Forensic Claims Consultant with AACE International. He is also a Fellow with the Project Management College of Scheduling. He has served in the construction industry for over 20 years, analyzing complex construction projects and providing expert testimony on scheduling, delays, productivity losses and damages. Brian is the co-author of the book Construction Delays: Understanding Them Clearly, analyzing them Correctly, and has authored a number of articles for construction industry publications. As someone who has worked on airport, bridge, commercial entertainment, government, healthcare high rise, highway and sports arena projects (in addition to others), Brian has seen both exceptional management strategies and a fair share of missed opportunities. Knowing that every project provides a tremendous learning experience for the next project, he seeks to share his experiences in this book, so the prudent professional will have the opportunity to learn from the past and improve upon their respective futures.

Index

Printed and bound by CPI Group (UK) Ltd, Croydon, CR0 4YY

26/03/2023

03205385-0001